The Open University

MU120
Open Mathematics

Unit 15

Repeating patterns

MU120 course units were produced by the following team:

Gaynor Arrowsmith (Course Manager)
Mike Crampin (Author)
Margaret Crowe (Course Manager)
Fergus Daly (Academic Editor)
Judith Daniels (Reader)
Chris Dillon (Author)
Judy Ekins (Chair and Author)
John Fauvel (Academic Editor)
Barrie Galpin (Author and Academic Editor)
Alan Graham (Author and Academic Editor)
Linda Hodgkinson (Author)
Gillian Iossif (Author)
Joyce Johnson (Reader)
Eric Love (Academic Editor)
Kevin McConway (Author)
David Pimm (Author and Academic Editor)
Karen Rex (Author)

Other contributions to the text were made by a number of Open University staff and students and others acting as consultants, developmental testers, critical readers and writers of draft material. The course team are extremely grateful for their time and effort.

The course units were put into production by the following:

Course Materials Production Unit (Faculty of Mathematics and Computing)

Martin Brazier (Graphic Designer)
Hannah Brunt (Graphic Designer)
Alison Cadle (TEXOpS Manager)
Jenny Chalmers (Publishing Editor)
Sue Dobson (Graphic Artist)
Roger Lowry (Publishing Editor)

Diane Mole (Graphic Designer)
Kate Richenburg (Publishing Editor)
John A.Taylor (Graphic Artist)
Howie Twiner (Graphic Artist)
Nazlin Vohra (Graphic Designer)
Steve Rycroft (Publishing Editor)

This publication forms part of an Open University course. Details of this and other Open University courses can be obtained from the Student Registration and Enquiry Service, The Open University, PO Box 197, Milton Keynes MK7 6BJ, United Kingdom: tel. +44 (0)845 300 6090, email general-enquiries@open.ac.uk

Alternatively, you may visit the Open University website at http://www.open.ac.uk where you can learn more about the wide range of courses and packs offered at all levels by The Open University.

To purchase a selection of Open University course materials visit http://www.ouw.co.uk, or contact Open University Worldwide, Walton Hall, Milton Keynes MK7 6AA, United Kingdom, for a brochure: tel. +44 (0)1908 858793, fax +44 (0)1908 858787, email ouw-customer-services@open.ac.uk

The Open University, Walton Hall, Milton Keynes, MK7 6AA.

First published 1996. Second edition 2004. Third edition 2008.

Copyright © 1996, 2004, 2008 The Open University

Edited, designed and typeset by The Open University, using the Open University TEX System.

Printed and bound in the United Kingdom by The Charlesworth Group, Wakefield.

ISBN 978 0 7492 2872 9

3.1

Contents

Study guide

This unit focuses on the use of sine curves to model periodic behaviour. It is intended to consolidate earlier work on the sine function from *Units 9, 13* and *14*, and also to provide more discussion of mathematical modelling.

The unit has three sections and includes some optional material.

Section 1 reviews the basic mathematical ideas associated with sines and their relationship to motion in a circle.

Section 2 looks at an example of modelling in which a sine curve is fitted to a set of data. The calculator is a superb tool in this context, and there is a section in the *Calculator Book* that guides you through the relevant modelling process. Section 2 also extends your library of trigonometric identities, and there is an optional section in the *Calculator Book* on this theme.

Section 3 aims to show how complex periodic variations, such as those associated with musical sounds, can be modelled by adding sine curves together. There is a video band as well as an audio band for this section. The video band, which is called 'The sound of silence', sets the context for the ideas covered in more detail in the rest of the section. You certainly need to view the video and undertake the associated activity (Section 3.1) before embarking on the main work of the section—you may find it best to deal with the rest of Section 3 in a separate study session.

In the audio band (Section 3.2), you will hear how a piano tuner can use musical beats to produce 'equally-tempered tuning'. Section 3.3 provides an introduction to Fourier analysis; it involves working through a section of the *Calculator Book* in which a useful calculator program is developed. Again, there is an optional section of the *Calculator Book* to study if you have time. Time will also determine whether you are able to take up the suggestion of watching the video again at the end of Section 3.

Throughout this unit you should add new mathematical ideas, formulas and identities to your previous notes on trigonometric functions. Also look for opportunities to add to the notes that you have made on modelling and regression.

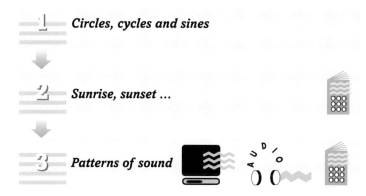

1   **Circles, cycles and sines**

2   **Sunrise, sunset ...**

3   **Patterns of sound**

Summary of sections and other course components needed for *Unit 15.*

Introduction

> What the mathematician does is examine abstract 'patterns'—numerical patterns, patterns of shape, patterns of motion, patterns of behaviour, and so on. Those patterns can be either real or imagined, visual or mental, static or dynamic, qualitative or quantitative, purely utilitarian or of little more than recreational interest. They can arise from the world around us, from the depths of space and time, or from the inner workings of the human mind.
>
> Keith Devlin (1994) *Mathematics: the science of patterns*,
> Scientific American Library, New York, p. 3

Repetitive events lie at the very root of human existence. Electrical brain rhythms, heart beats, sleeping and waking patterns—these are some of the regular biological activities that govern people's lives. Elsewhere, the relative periodic movements of the Earth, Sun and Moon determine the tides, the lengths of day and night, the monthly lunar cycle, the seasons, and the length of a year. Repeating patterns such as these form the backdrop against which people live their lives and, arguably, influence the way people think.

From a mathematician's point of view, seeing the world in terms of patterns offers a way of making sense of complicated behaviour—the behaviour can be thought of in terms of simpler mathematical 'building blocks' that are put together according to certain rules.

This unit focuses on the sine curve—a mathematical building block for a wide range of periodic behaviour. You met the sine curve in *Unit 9*, where it was used to model a sustained musical note from a tuning fork. You also saw in Chapter 9 of the *Calculator Book* how sine waves could be added together to produce other regular waveforms that were not themselves sine curves but that repeated periodically. In this unit, you are going to explore the mathematics of the sine curve further, and use the calculator to display the features of various periodic shapes.

1 Circles, cycles and sines

Aims   The main aim of this section is to review some of the mathematical language used to talk about sine curves and periodic behaviour.   ◇

You first met sine curves in *Unit 9* where they were used to model pure musical notes. There, and in *Unit 14*, you saw that there was a relationship between circles and the trigonometric functions, sine and cosine. Later in this unit, you will see that sums of sine and cosine functions offer a way of describing complex periodic behaviour. But what is so special about these particular trigonometric functions?

Activity 1   *Revising trigonometry*

Refresh your understanding of trigonometry by looking back at the relevant sections of *Units 9* and *14* and at any notes that you made on sines and cosines.

The key to understanding why sines and cosines are so closely linked to repetitive events lies in thinking about motion in a circle.

1.1 Exercise bike

Imagine you are sitting on an exercise bike. Your feet are pushing the pedals around and around at a steady rate. You are working hard—but not going anywhere! If you plotted the vertical height of one of the pedals above the pivot against the *horizontal displacement* of that pedal relative to the pivot, you would produce a circular graph, as in Figure 1.

Throughout this section, the height and horizontal displacement of the pedal are always measured relative to the central pivot.

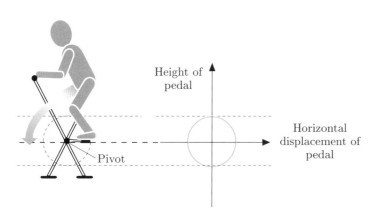

Figure 1   Pedalling an exercise bike: height of one of the pedals plotted against the horizontal displacement of that pedal.

7

▶ If you plotted the vertical height of one of the pedals against *time*, what shape of graph would you get?

Figure 2 shows the situation. Notice that this is a distance–time graph, a form of graph you first met in *Unit 7*.

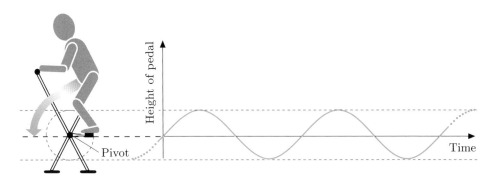

Figure 2   Pedalling an exercise bike: height of the pedal plotted against time.

Assume you are pedalling so that the pedals go around at a *steady rate* and are turning anticlockwise. Your foot reaches its highest point when the pedal is at the top of its circular path, and its lowest when the pedal is at the bottom. Thus, the pedal travels up and down, moving equally above and below the pivot each time around. When you plot the height of the pedal, relative to the pivot, against time, you get a graph whose shape you should recognize from earlier units.

The shape of the graph is a sine curve. In this case, the sine curve derives from the circular motion of the pedals. This is evident if you focus on the visible (left foot) pedal in Figure 2: as the pedal rotates at a steady rate, the shape of the curve repeats itself, producing the periodic, or regularly repeating, pattern of peaks and troughs, characteristic of a sine curve.

For each complete rotation of the pedal, the sine curve passes through one complete *cycle*, and then starts on another cycle. Look at Figure 3.

Note the use of the mathematical term 'cycle', and compare this with its everyday use.

Suppose the pedal is level with the pivot—at the three o'clock position, E in Figure 3. Since you are measuring the height of the pedal relative to the pivot, the height at E is zero, and therefore the sine curve passes through zero at that point. A quarter of a turn later, the pedal is at its highest point, F, and the sine curve is at its peak. After another quarter of a turn, the pedal is level with the pivot again—at the nine o'clock position, G—and the sine curve passes through zero again. The pedal then moves down, passes through its lowest point, H, and moves back up to the three o'clock starting position. Thus the lower half of the pedal's path mirrors the upper half. The sine curve correspondingly passes through a trough—its minimum value—and returns to zero having gone through one complete cycle.

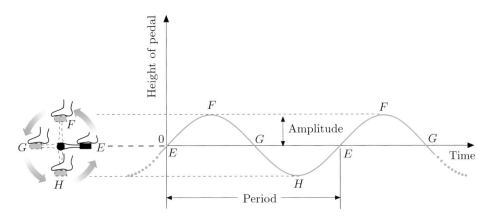

Figure 3   Period and amplitude of a sine curve.

It has been assumed that the turn started at point E, but, in fact, it does not matter where you start. You could start with the pedal anywhere on its circular path, and follow it all the way around through the highest and lowest points back to the starting position. Wherever you choose to start, as the pedal travels once around its circular path, the corresponding sine curve will go though one complete cycle. The shape of the curve is repeated for each complete turn, or rotation, of the pedal.

A turn is often referred to as a rotation, particularly in scientific and mathematical contexts.

Look at Figure 3 again. As you may recollect from *Unit 9*, the time required for one complete cycle of a sine curve is called the *period*. If you were pushing the pedal around one complete turn every second, the sine curve would have a period of 1 second. If you were pedalling faster and managing two complete turns every second, then the sine curve would go through two complete cycles in 1 second, giving a period of $1/2$ second.

The number of cycles that a sine curve goes through in one unit of time (usually 1 second) is called the *frequency*. Frequency and period are inversely proportional to one another. If the period is 3 seconds, then in 1 second the sine curve will trace out $1/3$ of a cycle: hence the frequency is $1/3$ cycle per second.

Recall inverse proportional relationships from *Unit 13*.

In general, a higher frequency means more cycles per second and consequently a shorter period. Conversely, a lower frequency means fewer cycles per second and a longer period. The relationship between frequency, f, and period, T, is

$$f = \frac{1}{T} \quad \text{or} \quad T = \frac{1}{f}.$$

Recall from *Unit 9* that the unit of frequency is the hertz (written Hz), where 1 Hz is 1 cycle per second.

The maximum deviation of a sine curve above or below the centre line (the horizontal axis) is the *amplitude*, A. Because the length of the pedal crank on a typical exercise bike is about 17 cm, the pedal reaches a maximum height of 17 cm above the pivot and a minimum height of 17 cm below the pivot. Therefore the amplitude of the corresponding sine curve is 17 cm.

To summarize:

- *amplitude* (A), plotted on the vertical axis, is the maximum height of the sine curve above or below the horizontal axis;
- *period* (T), plotted on the horizontal axis, is the time (often measured in seconds) required for one complete cycle;
- *frequency* (f) is the number of complete cycles per second (often measured in hertz);
- period and frequency are inversely related, so

$$T = \frac{1}{f} \text{ and } f = \frac{1}{T}.$$

Activity 2   *Reading the sines*

What is the amplitude, the period (in seconds) and the frequency (in Hz) of each of the three sine curves in Figure 4?

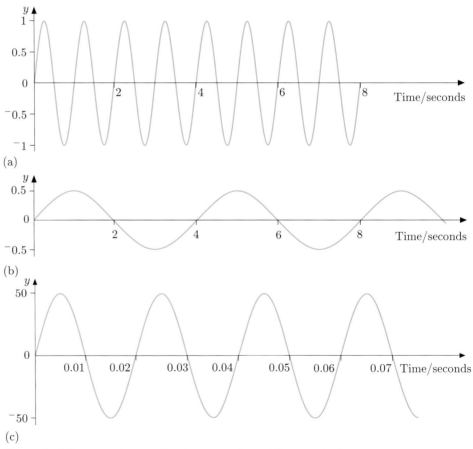

Figure 4   What are the amplitudes, periods and frequencies?

1.2 *Sine writing*

As you have just seen, the circular motion of either of the pedals of an exercise bike when pedalled at a steady rate produces, or 'generates', a sine curve. This curve is obtained by plotting the height of the pedal against *time*.

The same *shape* of curve can also be produced by plotting the height of the pedal against θ, *the angle turned through at the centre of the circle* (see Figure 5). In fact, sine curves associated with rotation are often drawn with angles plotted on the horizontal axis, and the unit of angle that is most frequently used is the radian, rather than the degree. From your previous work, you know that one complete turn, or once around a circle, is equal to $360°$, or 2π radians.

Activity 3 *Pieces of pi*

How many radians correspond to a quarter turn, a half turn and a three-quarter turn around a circle? Express your answers in terms of π.

In *Unit 9*, you saw that if you plot the function $y = \sin\theta$ on your calculator [actually entered as $Y = \sin(X)$], a sine curve like the one shown in Figure 5 will be produced. The curve has a maximum of 1 and a minimum of $^{-}1$, so the amplitude is 1. Notice that the horizontal axis shows the angle θ (in radians) rather than time as in Figure 3. As θ goes from 0 to 2π radians, corresponding to one complete turn around a circle, the graph of $y = \sin\theta$ goes through one cycle. Here the period of the sine curve is not measured in seconds, but in radians: the period is 2π radians.

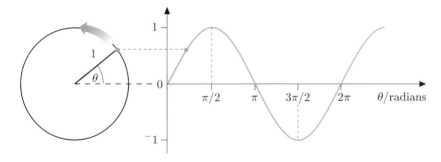

Figure 5   Generating the graph of $y = \sin\theta$.

There is an important fundamental difference between the graph in Figure 5 and the one in Figure 3. The graph in Figure 5 has been obtained by plotting a mathematical function rather than by plotting a series of observations as in the case of the exercise bike. It is the standard sine curve.

▶ How does the standard sine curve relate to the curve produced by observation? In other words, how would you use the sine function represented in Figure 5 to model the features of the curve produced in the exercise bike example in Figure 3?

The adjective 'sinusoidal' means 'like a sine curve'.

Take stock of what you know so far. For the exercise bike, the variation in the height of the pedal with time is sinusoidal with an amplitude of A and a period of T seconds, the actual value of T depending on how quickly the bike is being pedalled. In contrast, the standard sine curve in Figure 5 has an amplitude of 1 and a period of 2π radians. So you would need to make two changes to the standard curve in order to get it to match the actual curve: the amplitude must be scaled so that it matches the pedal amplitude, and the variable θ must be modified so that the two periods match.

All you need to do to change the amplitude from 1 to A is to *scale*, or multiply, the sine function by A. If y represents the variation in height of the exercise bike's pedal in centimetres, the changing height can be represented, or modelled, by the equation

$$y = A\sin\theta.$$

Next you need to modify θ in order to make the periods of the curves match. This will involve a number of steps. Start by linking the variable θ, the angle turned through (measured in radians), to the rate of rotation of the pedal. Look at Figure 6(a), which represents the variation in height of the pedal plotted against time for the case where the pedal completes one rotation every second: you can see that the period of rotation is 1 second, and that the period of the curve is 1 second. Figure 6(b) shows the variation when the pedal is going around twice as fast: here the period of rotation and the period of the curve are 0.5 second. Figure 6(c) shows the general case: on the horizontal axis, time is represented by the variable t, and T is used to represent the period of the curve and is also the time for one complete rotation of the pedal.

In Figure 6(d), the curve $y = A\sin\theta$, which models the motion of the pedal, is plotted. This curve has a period of 2π radians. If you compare (c) and (d), you should be able to see that there is a relationship between the way the variable t varies and the way the variable θ varies.

▶ Can you see what this relationship is?

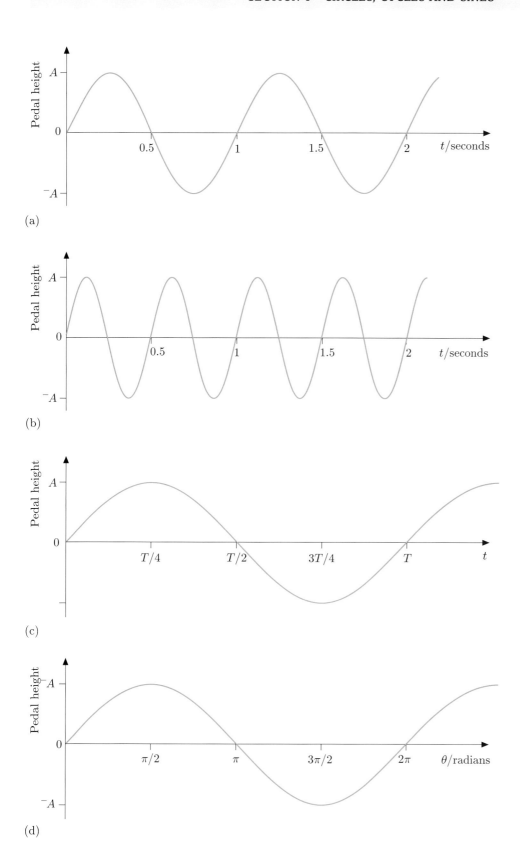

Figure 6 Sine curves of period (a) 1 second, (b) 0.5 second, (c) T seconds.
(d) The curve $y = A \sin \theta$.

By comparing the shapes of curves (c) and (d), you can see that $t = 0$ corresponds to $\theta = 0$, $t = T/4$ corresponds to $\theta = \pi/2$, $t = T/2$ corresponds to $\theta = \pi$, and so on.

Since the period T seconds corresponds to 2π radians, it follows that

$$1 \text{ second corresponds to } \frac{2\pi}{T} \text{ radians,}$$

and

$$t \text{ seconds correspond to } \frac{2\pi}{T}t \text{ radians.}$$

In other words,

$$\text{number of radians } \theta = \frac{2\pi}{T}t.$$

You can check this formula by substituting different values of t into it. At the start, $t = 0$, therefore θ is equal to $(2\pi/T) \times 0$, which is 0. When the pedal has gone halfway round, $t = T/2$, therefore $\theta = (2\pi/T) \times T/2 = \pi$. When the pedal has completed one turn, $t = T$, therefore $\theta = (2\pi/T) \times T = 2\pi$.

Recall that the height y of the pedal (relative to the pivot) can be modelled by the equation $y = A\sin\theta$. You can now replace θ by $\frac{2\pi}{T}t$ so that the equation gives the height of the pedal as a function of the time t. The resulting function,

$$y = A\sin\left(\frac{2\pi}{T}t\right),$$

is, of course, a sine curve and provides a mathematical model of the relationship between the pedal height y (measured in centimetres) and the time t (measured in seconds) for any given period of rotation T.

Thus the standard sine curve has been adapted to match the curve obtained for the exercise bike: the amplitude has been scaled and the period modified as required.

Example 1

Suppose that the length of the pedal crank is $17\,\text{cm}$ and that the period is $0.5\,\text{s}$.

(a) Sketch the variation of pedal height with time, assuming that the pedal starts from the three o'clock position.

(b) When does the pedal first reach its maximum height?

(c) What is the height of the pedal after $0.3\,\text{s}$?

Solution

(a) The amplitude A will be the length of the pedal crank, so $A = 17\,\text{cm}$. The period $T = 0.5\,\text{s}$, so $2\pi/T = 2\pi/0.5 = 4\pi$. Hence the equation representing the motion of the pedal is

$$y = A\sin\left(\frac{2\pi}{T}t\right) = 17\sin(4\pi t).$$

Figure 7 shows the variation of pedal height plotted against time.

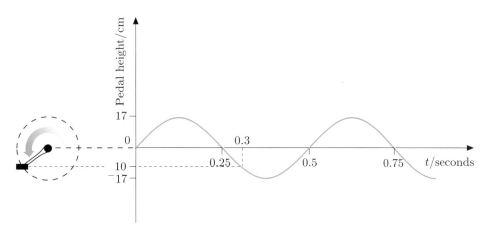

Figure 7   Height of the pedal plotted against time.

(b) The maximum height is first reached by the pedal when $y = 17\,\text{cm}$. This occurs after a quarter of a period; that is, when $t = T/4 = 0.5/4 = 0.125\,\text{s}$. You can check this by substituting 0.125 for t in the equation $y = 17\sin(4\pi t)$:

$$y = 17\sin(4\pi \times 0.125) = 17\sin(\tfrac{\pi}{2}) = 17\,\text{cm}.$$

(c) After $0.3\,\text{s}$, the height of the pedal is given by

$$y = 17\sin(4\pi \times 0.3) = 17\sin(1.2\pi) = {}^{-}9.99\,\text{cm (to 2 d.p.)}.$$

Therefore, after $0.3\,\text{s}$, the pedal is about $10\,\text{cm}$ below the pivot.

This position is shown in Figure 7.

It is often easier to talk about the *frequency* of a periodic event—how often it occurs in a given time (usually 1 second)—rather than the *period*—the time for just one cycle to occur. Recall from Section 1.1 that if f represents frequency in hertz and T represents the period in seconds, then

$$f = \frac{1}{T} \quad \text{or} \quad T = \frac{1}{f}.$$

So the formula for the sine curve that models the motion of the pedal can be expressed in terms of frequency rather than period. Start with

$$y = A\sin\left(\frac{2\pi}{T}t\right)$$

and replace $\dfrac{1}{T}$ by f, then the formula becomes

$$y = A\sin(2\pi f t).$$

Activity 4   Sketching the sine curve

A sine curve is used to model a periodic variation with a frequency of 3 Hz and an amplitude of 0.1.

(a) What is the formula that describes this curve?

(b) What is the period of the sine curve?

(c) Use your calculator to display this sine curve over the time interval $t = 0$ to $t = 1$ second, and so find when the first peak occurs. Sketch the result, remembering to mark the axes with appropriate labels.

Remember that frequency in hertz is a measure of the number of cycles completed every second, and that each complete cycle corresponds to 2π radians.

When the frequency f (in hertz) is multiplied by 2π, the resulting quantity, $2\pi f$, gives the number of radians turned through in each second. The quantity $2\pi f$ is called the *angular frequency*. It is measured in radians per second and is usually denoted by the symbol ω (the Greek letter 'omega'). Thus

$$\omega = 2\pi f.$$

It follows that if $f = 1\,\text{Hz}$, then the corresponding angular frequency ω is 2π, or about 6.28 radians per second. Similarly, a frequency of 50 Hz corresponds to an angular frequency of $50 \times 2\pi = 100\pi$, or 314.2 radians per second.

In science and technology as well as in mathematics, the formula for a sine curve is often expressed using angular frequency rather than frequency in hertz. You should be prepared to handle both. Consequently, the formula for the height of the exercise bike pedal can be written in another way: replacing $2\pi f$ by ω in $y = A\sin(2\pi f t)$ gives

$$y = A\sin\omega t.$$

Recall from *Units 7* and *11* the distinction between speed and velocity.

Since a radian is a measure of angular distance, the number of radians turned through per second, ω, is a measure of angular speed. When there is a direction involved (turning can be forwards or backwards) the term *angular velocity* is used. For example, riding the exercise bike so that the pedals turn once a second in the anticlockwise direction means that their angular velocity is 2π radians per second. An angular velocity of $^-2\pi$ radians per second would indicate that you were backpedalling at the same speed.

As you have just seen, there are two terms that you may come across to describe the quantity ω: angular frequency and angular velocity. Which one is used depends on the context. If you are dealing directly with things where the rotation has a physical meaning and, hence, the direction of rotation is relevant (as in the case of a wheel or a motor or a washing-machine drum), then angular velocity is the obvious choice. On the other hand, you may come across sine curves that are not associated with physical rotation (they may be modelling the sound of a musical note or an electrical signal, for example) or sine curves that are simply mathematical entities in their own right. Then, the term 'angular frequency' is more likely to be used.

There is a useful relationship between angular frequency ω and period T. Since ω is equal to $2\pi f$, the frequency f is $\omega/2\pi$. Now $T = 1/f$, so substituting for f gives

$$T = \frac{2\pi}{\omega}.$$

This equation says that ω is inversely proportional to T, with a constant of proportionality of 2π.

Multiplying both sides of this equation by ω gives the relationship $\omega T = 2\pi$. In other words, multiplying the angular frequency by the period (where both are in appropriate units such as seconds and radians per second, respectively) always gives the answer 2π. This constant relationship is useful in working out T if you know ω, or working out ω if you know T. It is also useful as a check on calculations: the value of ωT must always be equal to 2π.

Example 2 Middle C

A tuning fork sounding middle C vibrates 256 times per second, producing a pure note at a frequency of 256 Hz. What is the corresponding angular frequency? What is the period of the vibration?

Solution

The angular frequency $\omega = 2\pi f = 2\pi \times 256 = 1608.5$ radians per second.

When you work with the frequency in hertz, that is in cycles per second, you can calculate T using the relationship $T = \dfrac{1}{f}$. This gives

$$T = \frac{1}{256} = 0.00391 \, \text{s}.$$

So the tuning fork vibrates once in just under four-thousandths of a second.

Alternatively, if you work with the angular frequency in radians per second, you can use the relationship

$$T = \frac{2\pi}{\omega}.$$

Then

$$T = 2\pi/1608.5 = 0.00391 \, \text{s}.$$

As you would expect, the period comes out the same whichever way you carry out the calculation.

You have now seen that there are three alternative ways of writing the formula for a sine curve of amplitude A. They are as follows:

If you know the *period T*, the variation of y with time t is described by the formula

$$y = A \sin\left(\frac{2\pi}{T}t\right).$$

If you know the *frequency f* (in hertz), the formula can be written as

$$y = A \sin(2\pi f t)$$

If you know the *angular frequency ω*, the formula can be written as

$$y = A \sin(\omega t).$$

Activity 5 *Interpreting the formula*

The vertical position y (expressed in metres) of the needle of a sewing machine, when it is working at a certain setting, is described by the formula

$$y = 0.001 \sin(30t).$$

(a) What are the amplitude and the period of the needle's motion?

(b) Display two complete cycles of this sine curve on your calculator, and check how long the two cycles take.

(c) At a different setting, the amplitude of the needle's movement is the same, but the frequency is 8 cycles per second. What is the formula for y, the needle's position, in this case? Display this sine curve on your calculator, and compare it with the previous one.

In summary, a sine curve is a defined mathematical shape that can be used to model periodic behaviour. The curve is usually described by two parameters: the amplitude and the period (or frequency). The amplitude defines the 'height' of the curve—the maximum deviation of the peaks from the centre line. The period defines the 'width' of the curve—the time for one complete cycle (the frequency gives the number of cycles per second).

You have seen how a sine curve can be generated by a steady circular motion. The amplitude is equal to the radius of the circle described by the motion, and the period of the curve is equal to the time taken for one complete rotation or turn around the circle. The number of cycles completed in 1 second is called the frequency. Frequency is measured in hertz, where 1 Hz is equal to 1 cycle per second.

In this field of mathematics, angles and turning are usually specified using radians. One complete rotation, cycle or period is associated with turning through 2π radians. Hence the angular frequency ω, measured in radians per second, is related to the frequency in hertz by the formula $\omega = 2\pi f$.

Remember that the formula for a sine curve may be expressed using the amplitude A and either the period T, the frequency f, or the angular frequency, ω.

Activity 6   *Handbook activity*

Add the important points from this section to any previous notes that you have made on the sine function (see Activity 1).

In the next section you will undertake some modelling using the sine curve. As preparation, you should tackle Activity 7 which will help you to explore some simple transformations of the function $y = \sin(x)$.

Activity 7   *Tweaking the graph of $y = sin\,(x)$*

On your calculator, enter the standard function $y = \sin(x)$ and display the graph, using a suitable window setting. Also enter, in turn, each of the following functions, observing how its graph compares with the standard plot of $y = \sin(x)$:

$y = \sin(2x),$ $y = 2\sin(x),$

$y = 2 + \sin(x),$ $y = {}^-\sin(x),$

$y = 2 - \sin(x),$ $y = \sin(x) - 2,$

$y = 3 - 2\sin(x),$ $y = 3 - 2\sin(2x),$

$y = \sin(x - \frac{\pi}{4}),$ $y = \sin(x + \frac{\pi}{4}).$

Consider the general trigonometric function $y = a + b\sin(cx + d)$. What effect do the values a, b, c and d have on the position and shape of the curve relative to the standard graph $y = \sin(x)$?

You will have the opportunity to consolidate these ideas shortly, when working through Section 15.1 of the *Calculator Book* in the course of the next section.

Outcomes

After studying this section, you should be able to:

◇   use the following terms accurately and explain them: 'radian', 'amplitude', 'period', 'frequency', 'angular frequency', 'cycles per second', 'hertz' (Activities 3 and 6);

◇   describe in your own words the relationship between circular motion and a sine curve (Activities 1 and 6);

◇   explain and use the mathematical relationships between frequency, angular frequency and period (Activities 3, 4 and 5);

◇   write down the formula for a sine curve, given the amplitude and either the period, frequency or angular frequency (Activities 4, 5 and 6).

2 Sunrise, sunset ...

Aims   The main aim of this section is to illustrate the use of sine curves in modelling.   ◇

This section looks at how the annual repeating pattern of sunset times can be modelled using a mathematical formula. In the UK, it is part of everyday experience that the times at which the Sun rises and sets depend on both the time of year and the place. In London, for example, the midwinter Sun rises at about 08.00 GMT (Greenwich Mean Time) and sets just after 15.50 GMT. In Glasgow, about 480 km to the north and 260 km to the west of London, the midwinter Sun does not rise until after 08.45 GMT and sets around 15.45 GMT, making Glasgow's day in winter about 50 minutes shorter than London's. In midsummer, the situation is reversed; Glasgow's day starts at about 03.30 GMT, and ends about 21.00 GMT. London's midsummer day starts about ten minutes later and is about an hour shorter. From midwinter to midsummer and back again, the sunrise and sunset times at each location go through a pattern of values that repeats every year.

Some daily newspapers publish the predicted times of sunrise and sunset, and some diaries contain such information in the form of a table of places and times for each week of the year. Tables of data may be convenient ways of summarizing information and providing answers to *particular* questions such as, 'What time is sunrise today?', but such lists of numbers provide little insight into any *general* patterns or trends that may underlie the data. By contrast, plotting sunrise or sunset data on a graph gives a visual indication of how the times vary over the year, and thus is a useful first step towards representing the data by a formula.

▶   From your own experience, what shape of curve do you think you would get if you plotted the time of sunset in a given place over the course of a year?

2.1 A sine of the times

In the UK, the time of sunset governs 'lighting-up time', when street lighting is switched on and when vehicles must by law have their lights on. Figure 8 overleaf shows a graph of the times of sunset in London plotted over an interval that includes the whole of one particular year. The data points represent the time of sunset on the same day of the week in successive weeks (for example, on every Saturday). Week 0 represents the week from 6 January to 12 January, while week 51 represents the week from 28 December to 3 January the following year. To give you a better idea of the overall shape of the graph and how it repeats itself, the horizontal scale has been extended back from week 0 to week ⁻8 so as to include sunset data for the last eight weeks of the previous year, and it continues beyond week 51 to include data for the first few weeks of the following year.

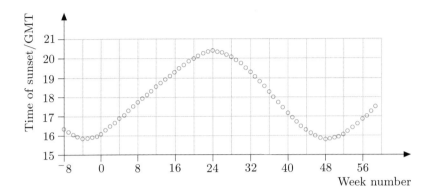

Figure 8   Variation in the time of sunset in London for a particular year.

As you can see in Figure 8, the time of sunset in London varies smoothly from 15.52 GMT in winter to 20.22 GMT in summer, a total range of 4.5 hours. Governed by the steady motions of the Earth, this cycle repeats itself (with very small variations) year after year. It is another example of periodic behaviour.

To predict the time of sunset for any day in a forthcoming year, astronomers use quite a complicated mathematical model that takes into account the exact position of the Earth in its orbit around the Sun. However, you can set up a simple but useful model of the periodic variation of sunset times, based on the data plotted in Figure 8 and the mathematics you have met so far.

Mathematical representations of data are used for a variety of reasons and come in a variety of forms. If you want a specific piece of information about an event, then it can simply be read off from a table or list. If you want to see whether there are any patterns or regularities in the data, then a graph may be the most appropriate way to present the information. But if you want to investigate more complex features of a set of data, it may be helpful to model the data using a suitable algebraic function.

In Block C, you used your calculator to find functions for modelling collections of data. You saw how you could fit straight lines, parabolas, exponential curves and power laws to data by employing regression techniques. In doing so, you were moving from graphical to algebraic models by matching a particular mathematical curve to the data, and thereby finding an algebraic function to represent the collection of data points.

▶   What kind of function would you choose to describe the annual pattern of sunset times?

By looking at the general shape of a set of data, you can often get an idea about what sort of curve, and hence what mathematical function, will fit the data, at least over the range you are interested in. Clearly, a straight line will not do to represent the sunset data—but what about other functions? The sunset curve is smooth and periodic—it repeats almost exactly year after year. Curves such as exponentials or parabolas may be made to fit different parts of a periodic curve with varying degrees of accuracy, but none has the property of smoothly repeating the same shape over and over again. The only functions you have met that do this are the trigonometric ones: sine and cosine. (Since the cosine curve is the same shape as the sine curve but shifted, or translated horizontally, you only need to consider the sine curve.)

The tangent is also a trigonometric function that is periodic, but it does not repeat smoothly. Use your calculator to plot it and see for yourself where the 'breaks' occur.

In fact, the variation in the time of sunset over a year looks promisingly like a sine curve—although it is not a standard sine curve. The peak of the curve is around week 24, coming about 27 weeks after the preceding midwinter trough and about 25 weeks before the following trough. Unlike a sine curve, therefore, the plot of sunset times is not quite symmetrical. Nevertheless, identifying the general shape is a good start.

So, choose a sine curve as the basic model and see how far you can get by using simple scalings and translations. In making this choice, you are stressing the features that the sunset curve has in common with a sine curve—its periodicity, its constant amplitude, its smooth variation from minimum to maximum and back to minimum over a cycle, and you are ignoring the fact that the sunset curve is not perfectly symmetrical like the sine curve.

Recall the sine curve model from Section 1:

$$y = A \sin\left(\frac{2\pi}{T}t\right),$$

where A is the amplitude of the sine curve and T is the period.

Activity 8 *Choosing the amplitude and period*

Use the graph of sunset times in Figure 8 to choose suitable values for the period T and the amplitude A in a sine curve model of these times.

If y represents the time of sunset in GMT (with hours expressed in decimal form) and t is the week number, then a first stab at writing down a model using the values for T and A identified in Activity 8 might give something like this:

$$y = 2.25 \sin\left(\frac{2\pi}{52}t\right).$$

Figure 9 overleaf shows the sunset times predicted by this *initial model* for various week numbers.

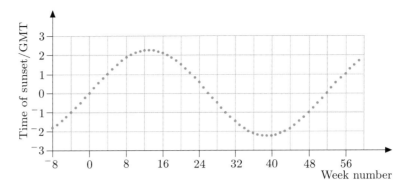

Figure 9   Plot of the predictions of the initial model, $y = 2.25 \sin\left(\frac{2\pi}{52}t\right)$.

The resulting curve has the correct amplitude of 2.25 hours and the correct period of 52 weeks, but if you compare it with the actual data in Figure 8 you can see that something is wrong: the curve has peaks in the wrong places, and the time of sunset has negative as well as positive values. However, these problems can be remedied.

▶   What must be done?

First, the curve must be shifted up so that it no longer varies above and below zero. Next, it must be shifted to the side so that the peaks and troughs are in the right places.

Over the year, the time of sunset in London varies from 15.52 GMT in midwinter to 20.22 GMT in midsummer. If the variation is assumed to be symmetrical, then the average time of sunset must be midway between these two times, at 18.07 GMT. When expressed as a decimal, 18 hours 7 minutes is 18.12 hours. Therefore, the curve must be shifted up so that it varies about 18.12, rather than about zero. To shift or *translate* the curve vertically as required, simply add the requisite amount to the formula:

Recall the discussion about translating lines and curves horizontally and vertically in *Units 10, 11* and *13.*

$$y = 18.12 + 2.25 \sin\left(\frac{2\pi}{52}t\right). \tag{1}$$

In Figure 10 the predicted sunset times from this *modified model* are plotted together with the actual data for comparison.

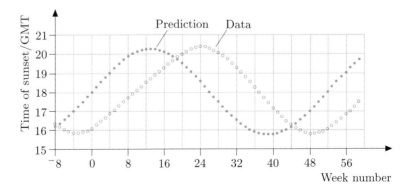

Figure 10   Plot of the sunset data and the predictions of the modified model, $y = 18.12 + 2.25 \sin(\frac{2\pi}{52}t)$.

Activity 9   *Displaying the predictions*

Enter equation (1) in your calculator and display the graph. Check for yourself that it has the shape shown in Figure 10.

The model now resembles the data more closely, but the peaks and troughs are still in the wrong places.

▶   What should be done to make the model an even better fit to the data?

To match the model to the data, the sine curve must be shifted, or translated, to the right. But by how much? Before tackling this question, look at Figure 11, which shows two sine curves. The black curve has the equation $y = \sin x$. At $x = 0$, the curve passes through 0, reaches its peak of +1 at $x = \pi/2$, returns to 0 at $x = \pi$, and so on. The green curve has the equation $y = \sin(x - \pi/2)$. It is the same in all respects as the black curve except it is translated by $\pi/2$ radians, or a quarter of a period, to the right relative to the first. So the second curve passes through ⁻1 at $x = 0$, then 0 at $x = \pi/2$, rises to ⁺1 at $x = \pi$, and so on.

Remember the period here is 2π radians.

The term 'phase shift' is often used in this context. Phase itself is a measure of how far a periodically recurring sequence of changes has progressed, and it is indicated graphically by the position on the horizontal axis. It follows that the shift between two sine curves (see Figure 11) is sometimes called a *phase shift*: thus $y = \sin(x - \pi/2)$ has a phase shift of $\pi/2$ radians, or a quarter of a period, to the right relative to $y = \sin x$.

Equivalently, you can talk of a phase shift of 90° instead of $\pi/2$ radians.

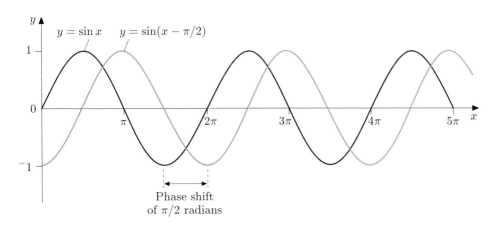

Figure 11   Phase shift of $\pi/2$ between two sine curves.

25

Activity 10   *Showing phase shifts*

(a) Use your calculator to display the curve $y = \sin(x + \pi/2)$. For which values of x between 0 and 2π is the curve equal to: 0, $^-1$, $+1$?

(b) Now display both $y = \sin x$ and $y = \sin(x + \pi/2)$. How could $y = \sin x$ be shifted to coincide with $y = \sin(x + \pi/2)$?

Shifting a sine curve by a quarter of a period to the right involves subtracting $\pi/2$ from x. Shifting it by a quarter of a period to the left, on the other hand, involves adding $\pi/2$ to x. Of course, you are not restricted to quarter-period shifts—a sine curve can be shifted left or right by adding or subtracting any appropriate amount. Adding $2\pi/3$ radians, for example, shifts a sine curve to the left by $2\pi/3$, or one-third of the period.

It is useful to remember that, in general, if the curve $y = \sin x$ has a phase shift of θ to the *left*, its equation becomes $y = \sin(x + \theta)$, and if the shift is to the *right*, the equation becomes $y = \sin(x - \theta)$.

Now return to the problem of improving the model of sunset times. Look at Figure 10 again and notice that there is a phase shift between the nearly sinusoidal curve of the sunset data and the sine curve produced by the modified model. The model predicts that the time of sunset during week 0, represented by $t = 0$, will be 18.12 hours (18.07 GMT). However, the data show that the Sun does not actually set at this time until 10 weeks later. For the model to match the data, the sine curve must be shifted to the right by an amount corresponding to 10 weeks. Since the period of the sunset data is 52 weeks, the shift must be 10/52 of a period, corresponding to $(10/52) \times 2\pi = 1.21$ radians. Incorporating this shift in the model produces the revised formula:

$$y = 18.12 + 2.25 \sin\left(\frac{2\pi}{52}t - 1.21\right). \tag{2}$$

This formula represents an *improved model* of the sunset times over a year. But how well does it fit the data? To answer this question you need to compare the predictions of this model with the data.

In Figure 12, the predictions of the improved model are plotted on the same graph as the sunset data to show where the values differ. The model is quite successful at giving the overall shape. However, as you can see, the sine curve generated by the model follows a slightly different path from the curve described by the data. The curves start together around week $^-3$, corresponding to mid-December. Until week 10, the curve due to the model underestimates the data, predicting sunset times that are slightly too early for this part of the year. The curves then cross each other just before the spring equinox, and thereafter the model overestimates the data, giving sunset times that are slightly too late, until week 23 (just before midsummer) when the model peaks slightly ahead of the data. For the second half of the year, the model consistently gives sunset times that are very slightly too early. The model and data coincide again around midwinter, and the cycle then starts again.

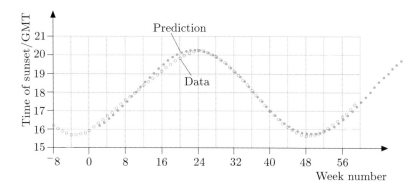

Figure 12   Plot of the sunset data and the predictions of the improved model, $y = 18.12 + 2.25 \sin(\frac{2\pi}{52}t - 1.21)$.

Example 3   *May Day*

The first day of May is the traditional day for springtime festivities. Use the improved model [equation (2)] to predict the time that the Sun sets in London at the start of the week in which May Day falls in this particular year.

Solution

May Day occurs in week 16 of the year. Putting $t = 16$ in the formula for the improved model

$$y = 18.12 + 2.25 \sin\left(\frac{2\pi}{52}t - 1.21\right)$$

gives

$$\text{sunset time} = 18.12 + 2.25 \sin\left(\frac{2\pi}{52} \times 16 - 1.21\right)$$
$$= 18.12 + 2.25 \sin(1.93 - 1.21)$$
$$= 18.12 + 2.25 \sin(0.72)$$
$$= 18.12 + (2.25 \times 0.66)$$
$$= 19.61, \text{ or } 19.37 \text{ GMT.}$$

The actual time of sunset for the start of this week is 19.18 GMT, so the model overestimates the time by 19 minutes. As you will see shortly, it turns out that the predictions of the model are most inaccurate around this time of year.

Activity 11   *Remember, remember . . .*

On 5 November in the UK, bonfires and fireworks commemorate Guy Fawkes' 'Gunpowder Plot', an attempt to blow up King James I and Parliament in 1605. Week 43 includes 5 November. Use the improved model to predict when the Sun sets in London at the start of the week in which Guy Fawkes' Night falls for the year in question.

An alternative way of assessing how well the improved model's predictions fit the actual data is illustrated in Figure 13. This is a plot of the difference in minutes between the predicted and the actual values of the sunset time during the course of the year. The difference is calculated as 'predicted' minus 'actual'.

A positive difference indicates that the prediction of the model overestimates the actual sunset time—the predicted time is too late—and a negative difference indicates that the model underestimates the data—the predicted time is too early.

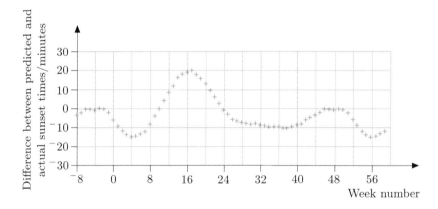

Figure 13   The difference between the predictions of the improved model and the actual sunset data.

The advantage of a difference plot like the one in Figure 13 is that it makes it easier to see where the match between the predicted and the actual data is good, and where it is less so. You can see that the difference, or error, around week ⁻4 is close to zero indicating a good match. Around week 4 the error reaches ⁻15 minutes, before passing through zero at week 10, and then reaching a maximum of 20 minutes around week 17; here, the match between model and data is poorest. The error returns to zero around week 24 and then does not exceed ⁻10 minutes for the rest of the year.

Achieving numerical accuracy is an important aspect of many mathematical models. In this case, the greatest difference between the predictions of the model and the actual data is nowhere more than 20 minutes for a situation where the annual variation in the data ranges over 4.5 hours. In other words, in the worst case the improved sine curve model predicts the time of sunset with an error of less than 8%. For a lot of the time, the match is better than this: for 32 consecutive weeks, the error does not exceed 10 minutes, or 4%. It is up to the user of a mathematical model to judge whether the predictions offer an acceptable fit to the data. Decisions depend on what the model is for and on the criteria for accuracy.

To talk of errors in percentage terms involves making a *relative* comparison, (as opposed to an *absolute* comparison). This idea was explained in *Unit 2*.

The differences between the predicted and real curves may also prompt questions about *why* they are different. Does the model ignore some features that should have been taken into account? Is there a good reason why the pattern of sunset times is not a perfect sine curve? Thinking about questions like these can give insight not only into how the model may be improved but also into the mechanism of the physical phenomenon itself.

The aim in this section has *not* been to produce a highly accurate model for predicting sunset times but, rather, to explore the modelling process more generally. However, if you were looking for greater numerical accuracy (less than 4% error over the whole year, say), then the next stage would be to think about how the model might be further modified to achieve the desired result. It may be that modifying the parameters of the model slightly will reduce the error or spread it more evenly over the year. Or it may turn out that a better result simply cannot be achieved with a model of this type. It may be obvious which changes to make or it may not be—there are no hard and fast rules. Producing an acceptable mathematical model in a particular situation requires a combination of mathematical skill, intuition and experience. In short, modelling is something of an art.

Parameters are the numbers in a model that represent quantities such as amplitude and phase shift.

The modelling process that has been used here to produce the improved model [equation (2)] can be summarized as follows:

◇   the data points were plotted and showed a periodic variation;

◇   the shape of the plot was characterized as a sine curve;

◇   the amplitude and the period were estimated from the plot;

◇   the sine curve was translated vertically by adding a constant;

◇   the sine curve was translated horizontally by including a phase shift;

◇   at each stage, the match between the model and the sunset data was checked, and a decision was made about what to do next;

◇   finally, the model was evaluated, the assumptions clarified, and possible improvements considered.

It is sometimes useful to express a model in general terms. Here the general form of the improved model of the sunset data is

$$y = M + A \sin\left(\frac{2\pi}{T}t + \phi\right),$$

where y represents the time of sunset, t is time of year (in weeks), M is the mean or average value of the sunset time, A is the amplitude of the sinusoidal variation, T is the period, and ϕ (the Greek letter 'phi') is the phase shift (positive or negative).

Activity 12 *Thinking about modelling*

Take some time to think about the modelling process that you have just been through. What assumptions have been made? Can you suggest any ways in which the model could be improved further so as to fit the data better? How would you define 'a better fit'?

Make some notes on the cream-coloured activity sheet, entitled 'Thinking about modelling'.

Next you will tackle a modelling problem with the aid of your calculator, and this should clarify and consolidate your understanding of the modelling procedure that has been introduced in this section.

Now work through Section 15.1 of the Calculator Book.

2.2 *Identity parade*

So far you have been using the sine function to model periodic behaviour. But this is not the only function available to mathematicians. Very closely related to the sine function is the cosine function, written $\cos x$. As the value of x increases from 0 to 2π, the value of $\cos x$ varies from $+1$ to $^-1$ and back again. Plotting the value of $\cos x$ against the value of x gives the cosine curve.

Activity 13 *Relating sines and cosines*

On your calculator, display a sine curve (the graph of the function $y = \sin x$) and a cosine curve (the graph of the function $y = \cos x$).

How are the curves related to each other?

You can think of the *cosine* curve as a sine curve that has been shifted to the left or right. For example, a simple shift would be to the left along the x-axis by an amount equal to $\pi/2$ radians, as in Figure 14(a); that is, a phase shift of a quarter of a period.

Perhaps make up a mnemonic (a memory aid) to help you remember which way the curves are shifted—maybe something like '**R**emoving $\pi/2$ means moving to the **R**ight'.

Recall from Section 2.1 that shifting a sine (or cosine) curve to the left involves adding to the angle, while shifting a sine (or cosine) curve to the right involves subtracting from the angle. So, the relationship between the cosine and sine curves in Figure 14 can be expressed mathematically by the equation

$$\cos x = \sin\left(x + \frac{\pi}{2}\right).$$

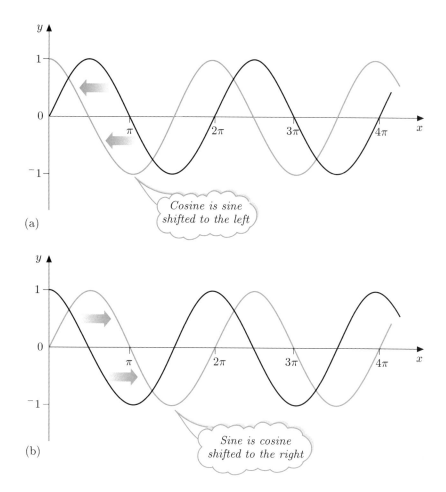

Figure 14   Shifting a sine curve and a cosine curve by $\pi/2$.

▶ How is the expression at the foot of the previous page to be interpreted?

All it is saying is that, for any value of x, the corresponding values of $\cos x$ and $\sin(x + \pi/2)$ are the same. For instance, if $x = 0.5$, then $\cos 0.5 = 0.8776$, and $\sin(0.5 + \pi/2) = \sin(2.0708) = 0.8776$. Try some other values for yourself.

You saw in *Unit 14* that there are many special relationships among the trigonometric functions, and these were described as identities. The relationship between $\cos x$ and $\sin x$ is another example of an *identity*, because it is true for all values of x. Therefore, for any value of x, the values of the cosine function and the shifted sine function are always the same.

You can also think of a *sine* curve as a cosine curve that has been shifted to the *right* (rather than to the left) along the x-axis by $\pi/2$ radians, or a quarter of a period, as in Figure 14(b). Hence the relationship between the sine curve and the shifted cosine curve is

$$\sin x = \cos\left(x - \frac{\pi}{2}\right).$$

This is another identity—an expression that is true for all values of x.

Identities	Equations
Examples: $2x + 6 = 2(x + 3)$ and $x^2 - 6x + 9 = (x - 3)^2$.	Examples: $2x + 6 = 18$ and $x^2 - 6x + 9 = 0$.
True for *all* values of x.	True only for *particular* values of x.
Cannot be solved.	Can (often) be solved.

Activity 14   *From cosine to sine*

Display the curve $y = \cos(x + \pi/2)$ on your calculator. How could you express this curve in terms of $\sin x$?

Shifting $\sin x$ or $\cos x$ by even multiples of π, that is by $2\pi, 4\pi, 6\pi, \ldots$, simply gives you the same function again. This is because the sine and the cosine are both periodic functions with a period of 2π. Look at Figure 15, which shows the situation for a sine curve.

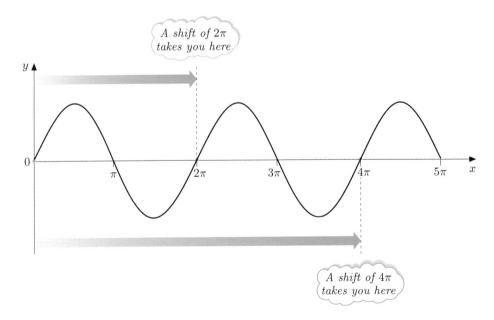

Figure 15   Shifting a sine curve by even multiples of π.

You can see in Figure 15 that moving through one or more complete periods (in either direction) takes you from your starting point to an identical y-value on the curve. This means that $\sin(x + 2\pi)$ and $\sin(x - 2\pi)$ are equal to $\sin x$, as are $\sin(x + 4\pi)$ and $\sin(x - 4\pi)$, $\sin(x + 6\pi)$ and $\sin(x - 6\pi)$, and so on. Writing this relationship as an identity gives

$$\sin x = \sin(x \pm 2\pi) = \sin(x \pm 4\pi) = \sin(x \pm 6\pi)\ldots$$

Note the use of the symbol $\pm$ to stand for 'plus or minus'.

or, more generally,

$$\sin x = \sin(x \pm n\pi) \text{ when } n \text{ is an even number.}$$

Shifting by odd multiples of π, that is by $\pi, 3\pi, 5\pi, \ldots$, effectively, turns the curve upside down, or 'inverts' it. Figure 16 shows that the peaks and troughs of $\sin(x + \pi)$ are the opposites of those of $\sin x$. The same is true for $\sin(x - \pi)$, $\sin(x + 3\pi)$, $\sin(x - 3\pi)$, $\sin(x + 5\pi)$, $\sin(x - 5\pi)$, and so on. So, written as an identity,

$$\sin x = {}^{-}\sin(x \pm \pi) = {}^{-}\sin(x \pm 3\pi) = {}^{-}\sin(x \pm 5\pi)\ldots$$

or, more generally,

$$\sin x = {}^{-}\sin(x \pm n\pi) \text{ when } n \text{ is an odd number.}$$

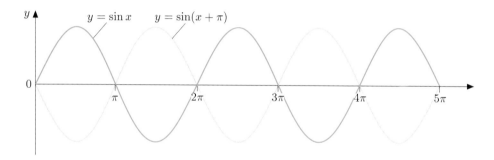

Figure 16   Shifting a sine curve by odd multiples of π.

These identities are also true for cosines because sines and cosines are just phase-shifted versions of each other. Therefore

$$\cos x = \cos(x \pm n\pi) \text{ when } n \text{ is an even number,}$$

and

$$\cos x = {}^{-}\cos(x \pm n\pi) \text{ when } n \text{ is an odd number.}$$

Try out these identities on your calculator with different integer values of n, and confirm for yourself the statement about odd and even values of n. Remember that in the case of an identity, the curves arising from the expressions on each side of the equals sign will be identical. Your calculator will draw one curve on top of the other, so if you have an identity *you will see what appears to be a single curve.*

Do not try to remember all these relationships. Instead work them out as necessary. You can do this by drawing a sketch or using your calculator and thinking about what you must do to one side of the identity to make it equal to the other. The next two activities give you practice at this.

Activity 15   *In a simpler form*

Below are four expressions, all of which can be written more simply in terms of either $\sin x$ or $\cos x$. Use your calculator to display corresponding functions and hence complete the table of sine and cosine identities. The first has been done for you.

Expression	Simpler form	Comment
$\sin(x - \pi/2)$	$^-\cos x$	$\sin x$ has been shifted to the right by $\pi/2$ radians.
$\cos(x + 2\pi)$		
$\sin(x - \pi)$		
$\cos(x + 3\pi/2)$		

You may not have the same answers for this activity as given in the Comments section—but that does not necessarily mean you are wrong. Because sine and cosine are both periodic functions, and because the only difference between them is a $\pi/2$ phase shift, there are any number of correct ways of expressing them. Sometimes the reason for expressing a sine or cosine function in a different form is to be able to write it more simply. Sometimes it is because it is easier to work just with sines or just with cosines, and not have a mixture of both.

The important point, however, is not that you worry about all the different ways that sines and cosines can be expressed, but that you practise displaying the curves on your calculator and so get used to recognizing the effects of phase shifts on the relative positions of the peaks and troughs.

Activity 16   *From sine to cosine*

Recall that the improved model for the sunset time is given by

$$y = 18.12 + 2.25 \sin\left(\frac{2\pi}{52}t - 1.21\right).$$

Use the last identity found in Activity 15 to express this model in terms of a cosine function instead of a sine function.

Activity 17   *Identity check*

In *Unit 14*, you saw that the application of Pythagoras' theorem to a right-angled triangle with a hypotenuse that is 1 unit in length led to the identity $\sin^2 \theta + \cos^2 \theta = 1$.

Use your calculator to convince yourself that this identity is true for any value of θ by displaying the functions $y = \sin^2 \theta$, $y = \cos^2 \theta$ and $y = \sin^2 \theta + \cos^2 \theta$ together.

In Section 2.1 you used the mathematics of sine curves to set up a mathematical model of the variation of the time of sunset over a year. The values of the parameters in the model were chosen so that the sine curve matched the available data.

In Section 2.2 you saw that sine and cosine curves are very closely related. Formulas using sines and cosines can be expressed in many different ways by means of various mathematical identities. A central idea is that $\sin x$ and $\cos x$ are identical except for a phase shift of a quarter of a period, or $\pi/2$ radians.

It is worth taking some time at this point to reflect more generally on mathematical representations, in particular those that are graphical or algebraic. Graphs and algebraic functions can both be used to model events in the real world. Graphs are useful because they store information conveniently in the form of points and lines, and they give a visual representation of how quantities change. You can learn to interpret the resulting shapes by sight, and make judgements based on the shape that the data present. Mathematical functions are useful because they can be used for detailed calculation and can be manipulated according to the rules of algebra. Functions are a concise way of conveying information, replacing graphical shapes by symbolic relationships.

Going from a graphical to an algebraic model of the times of sunset involved changing from one form of mathematical representation to another. Different representations imply different viewpoints—a different type of model represents a situation in an alternative way. A graphical model stresses relationships in geometric terms, telling a story using shape, size and position. An algebraic model stresses relationships in terms of symbols, telling a story using numbers, letters and mathematical functions.

A graph is usually expressed in the form

$y = $ some expression in x.

This form of expression directly links the horizontal and vertical coordinates.

An alternative approach is to define a graph by *two* separate statements, one giving the horizontal coordinate and the other giving the vertical coordinate. This is the basis of a technique called *parametric graphing*, which is explained in an optional section of the *Calculator Book*.

If you wish to learn about parametric graphing and see how it can be used to produce a range of interesting shapes on your calculator, work through Section 15.2 of the Calculator Book.

Activity 18   *Keeping up with your learning*

So far, this unit has involved you in revisiting ideas that you have met earlier in the course, and then developing and integrating those ideas into new work.

How did you check your understanding of earlier work?

How did you get on with using the mathematical terms?

Have you extended or revised your earlier notes on trigonometric functions? If not, add the ideas from this section to consolidate your work on *Units 9* and *14*, as well as on this unit.

Outcomes

After studying this section, you should be able to:

◇   use the following terms accurately and explain them: 'periodic behaviour', 'phase shift', 'identity' (Activities 10, 13, 15 and 18);

◇   choose appropriate values for the amplitude, period or frequency and phase shift of a sine curve model, given a set of periodic data (Activities 8 and 11);

◇   outline the modelling process involved in fitting a sine curve to a set of periodic data (Activity 12);

◇   explain how to translate a sine curve or cosine curve vertically or horizontally, by adding an appropriate constant or a phase shift (Activities 10 and 13);

◇   model data using sine regression on your calculator;

◇   relate the sine and cosine functions to each other, using standard trigonometric identities (Activities 10, 13, 14, 15, 16 and 17);

◇   interpret and evaluate an expression of the form $M + A\sin(\omega t + \phi)$ (Activities 9, 10, 11, 14 and 15);

◇   use your calculator to explore parametric graphing (optional).

3 Patterns of sound

Aims The main aim of this section is to show how complex periodic variations can be described by adding sine curves together. ◇

So far you have come across sine curves being used to model periodic behaviour that varies smoothly, and also sine curves as mathematical objects in their own right. This section looks at sine curves as building blocks for modelling sounds.

An important idea in technology and science, as well as in mathematics, is that many complex repeating patterns can be built up by adding together sine curves of different amplitudes, frequencies and phases. Such patterns can look quite different from individual sine or cosine curves, but it turns out that almost any periodic shape can be built up by adding together the right sinusoidal functions. The idea of adding sine curves was first introduced in Section 9.2 of the *Calculator Book*, and it may be helpful to glance back at that section now.

The power of this approach lies not only in being able to build up periodic shapes but also in being able to break them down into a collection of component sine curves. Consequently this view offers a new way of talking and thinking about periodic behaviour.

Sine curves, sine waves, sinusoids and waveforms

In mathematics, the term 'sine curve' is used to mean the graph produced by a function such as $y = \sin x$ or $y = \sin 3x$. Similarly, the term 'cosine curve' refers to the graph produced by a function such as $y = \cos x$ or $y = \cos 4x$. In these expressions, x is the independent variable. Remember, sine and cosine curves are identical in shape. Anything expressed as a sine function can also be expressed as a cosine function with a phase shift of $\pi/2$ radians or $90°$, and vice versa.

If you study science and technology, you are likely to come across the term *sine wave*, meaning a sine-like variation with time. For example, the variation in air pressure with time produced by the vibrations of a tuning fork can be modelled as a sine wave. A sine wave is represented by a function such as $y = \sin 2t$ or, more generally, as $y = A\sin(\omega t + \phi)$, where the variable t represents time. Sometimes, the general term *sinusoid* is used to indicate this type of function, whether it relates to a sine or a cosine variation with time.

The term *waveform* is also common in science and technology. It refers to the general shape of a variation with time. For instance, you may recall that the video associated with *Unit 9* showed oscilloscope displays of waveforms associated with musical sounds.

A word about notation. In the rest of this section, and more widely in mathematics and science, you will find expressions such as $\sin 2x$ being used, rather than the equivalent form, $\sin(2x)$. Both forms are acceptable, though the course calculator only allows you to use the bracketted form. Brackets are definitely necessary in expressions such as $A\sin(\omega t + \phi)$ and in cases where omitting them would lead to ambiguity.

3.1 The sound of silence

You should now read the following notes and then watch the video band 'The sound of silence'.

The video uses the context of musical notes to review some of the ideas about sine curves that you have already met here and in *Units 9* and *14*. It also looks forward and introduces a number of ideas that may be quite new to you. These ideas are discussed in more detail later in this section, so do not worry if you fail to understand everything the first time you watch the video. A good learning strategy is to watch all of the video initially, in order to review the basic concepts and get a flavour of what is to come. Then, if you have time when you have completed Section 3, you should watch the video again.

'The sound of silence' illustrates how a sine curve is related to the vertical displacement of the prongs of a tuning fork over time. It also shows two ways of representing a musical note graphically—the more familiar *waveform*, as well as a *frequency spectrum*.

Now watch band 12 of DVD00107 and carry out Activity 19.

Activity 19 Video notes

As you watch the video, make some notes and sketches so that you could explain the following ideas to someone who is not taking MU120. In particular:

(a) What is the relationship between vertical displacement over time and the shape of a sine curve? Explain the terms 'amplitude', 'period', 'frequency', 'phase shift', 'waveform' and 'frequency spectrum'.

(b) How can you produce silence by playing two notes?

The video has covered a number of different but related topics. As part of the process of consolidating your knowledge, you may find it helpful to reflect further on some of the main points of the video before you tackle the rest of the section. These points are summarized below.

The video explored the complex sounds produced by musical instruments. Different instruments playing the same note produce sounds at the same pitch but with a different quality, or timbre. The sound produced by a saxophone, for instance, is different from that of a flute, even though they may be playing the same note. Differences in timbre between instruments can be described by treating each sound as if it were made up of many different sine waves at different frequencies and amplitudes. Thus, when a note is played on a saxophone, what you hear is a blend of many harmonics, each with its own loudness (corresponding to the amplitude of its sine curve) and each with its own pitch (measured by its frequency).

As you saw in the video, it is possible to use a computer to produce a so-called frequency spectrum graph. This consists of a series of vertical lines, each representing one of the component sine waves. In the spectrum, the frequency of each component is shown on the horizontal axis, with the corresponding amplitude on the vertical axis. So, if you select a particular frequency on the horizontal axis, the height of the vertical line at that point is a measure, in terms of loudness, of how much of that component note is present in the overall sound that you hear. Each musical instrument has its own characteristic frequency spectrum, and these graphs will be discussed in more detail in Section 3.4.

Another topic considered in the video was how two notes, represented by sine waves, can be added together to produce silence. The background information for this topic was covered in Section 2.2, where you saw how a phase shift of π radians, or 180°, effectively turns a sine curve upside down, or inverts it. Shifting all the sine wave components of a sound by π radians has the effect of turning the entire periodic waveform upside down. When an inverted wave such as this is added to the original, the result is zero.

Sound-cancellation techniques based on similar principles can be used to reduce the level of noise from machinery in industrial environments.

So, two sounds can be added together to produce nothing—quite literally the sound of silence. Where the wave pattern is more complex, a simple phase shift will not invert the wave and therefore such a shift does not cancel the original sound. However, the result can still be achieved by inverting the wave, but in this case more sophisticated means have to be used to obtain the inversion.

You also saw (and heard) in the video how a sound is altered when the higher-frequency components are removed. The human voice can be thought of as being made up of a range of vocal sounds at different frequencies. Removing the higher-frequency components changes the characteristics of a voice. You may have come across this effect on the telephone. Voices can sound 'restricted' and 'tinny', and in extreme cases you may not be able to recognize the other person's voice at all. The video demonstrated this effect, and it will be discussed further in Section 3.3.

3.2 Adding sounds

In *Unit 9*, you saw how modern musical instruments are usually tuned according to the principle of equal temperament. This produces an equally-tempered scale, in which the intervals between the semitones are such that the frequencies of adjacent semitones are in the same ratio throughout; that ratio has a value of $\sqrt[12]{\left(\frac{1}{2}\right)}$, or approximately 0.9439.

Equal temperament is a convenient compromise that allows various musical instruments to be played in tune with each other in any key. But, to a trained musician's ear, such tuning creates notes that are not perfectly in tune because it fails to produce intervals based on very simple 'perfect' fractions like $\frac{2}{3}$ and $\frac{3}{4}$. As you will hear in the following audio section, this creates a problem for piano tuner Henry Tracy. He explains how he initially creates perfect Pythagorean intervals—that is, perfect octaves, perfect fourths and perfect fifths. Then, in order to create the compromise tuning of equal temperament, he has to exploit the phenomenon of 'beats'. In mathematical terms, beats occur when two sine waves with frequencies within a few hertz of each other are added together, giving another periodic wave whose amplitude also rises and falls regularly. The resulting regular amplitude variations are called 'beats'. Counting the number of beats that each pair of strings produces helps Henry gauge just how far he needs to tweak the tuning from the 'perfection' of Pythagorean tuning to meet the requirements of equal-temperament tuning.

Octaves and fifths are defined in the Glossary of *Unit 9*.

Activity 20 *Tuning a piano*

Listen to band 3 of CDA5570 (Tracks 8–9), 'Tuning a piano', and make notes on the stages that the piano tuner, Henry Tracy, goes through. As you listen, keep in view the diagram and photographs in the frames that follow.

Do not worry if you cannot clearly identify the beats—many people cannot hear them until trained to do so.

Frame 1

Henry Tracy tuning the piano.

Most notes are made by the hammers striking three strings.

Tuning pins for each string, three for each note.

Tuning fork for the note C above middle C.

Pulling the string up in pitch with the tuning key.

The bearing scale from F# below middle C (Henry's little finger) to F above middle C (Henry's thumb).

Frame 2

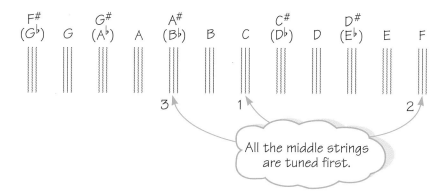

All the middle strings are tuned first.

Three strings for each note in the bearing scale.

The sequence of notes that Henry tunes in what he calls the bearing scale is

start with tuning middle C,
up a fourth (5 semitones) to tune F,
down a fifth (7 semitones) to tune A# (= B♭),
up a fourth (5 semitones) to tune D# (= E♭),
down a fifth (7 semitones) to tune G# (= A♭),
and so on.

Felt device blocking off the two outside strings for each note in the bearing scale.

A felt wedge blocking off one string of the three for a note so that the other two can be compared.

In *Unit 9*, it was shown how a pure musical tone can be represented by a sine function. When two such tones are sounded together, their combined effect can be modelled by *adding* the corresponding sine functions.

▶   So what is going on at a more detailed level when two such sine functions are added together?

Frequencies of 6 and 7 Hz are rather too low to hear, but are easier to show in a diagram.

Take the example of two sine waves representing pure tones of frequencies 6 Hz and 7 Hz, respectively, and with the same amplitude (see Figure 17). Over a time interval of 1 second, the 6 Hz sine wave goes through six cycles and the 7 Hz sine wave goes through seven cycles.

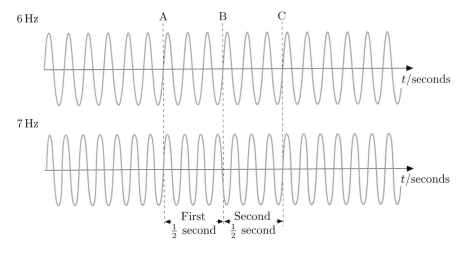

Figure 17   Sine waves of frequencies 6 Hz and 7 Hz moving in and out of phase.

Now, when two waves are at a point where they share the same position and are moving in the same direction, they are said to be in phase with each other. A specific example is shown at position A in Figure 17, where both waves are at zero and have positive gradients. As time increases, these waves gradually move out of phase with each other. After $\frac{1}{2}$ second, the 6 Hz wave has completed three full cycles and is back at zero with a positive gradient. However, the 7 Hz wave has completed $3\frac{1}{2}$ cycles, and so it is at zero but with a negative gradient. The waves are now completely out of phase (position B on Figure 17). Over the next $\frac{1}{2}$ second, the curves move back into phase again, passing through zero with positive gradients (position C). Thus the waves move in and out of phase every second.

Figure 18(a) shows the same two sine waves, and Figure 18(b) shows the result of adding these two waves together. The shaded portion of Figure 18(b) is shown using an enlarged scale in Figure 19.

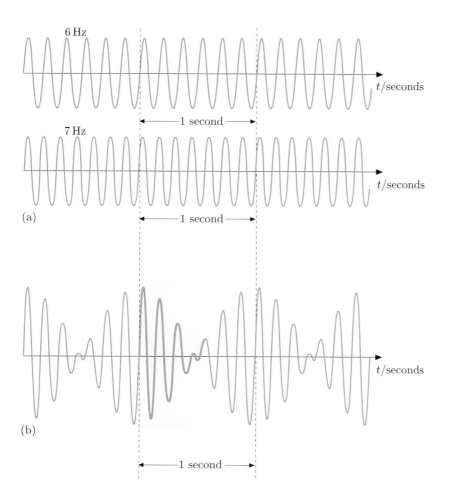

Notice that the sum of the two waves is twice the height of the individual waves when those waves are completely in phase.

Figure 18   (a) Sine waves of frequencies 6 Hz and 7 Hz, and (b) their sum.

Figure 19 indicates in graphical terms how the two sine waves combine. When both waves are positive, their sum is also positive; similarly, when both are negative, their sum is negative. But, when one is positive and the other is negative, the waves partly cancel each other out, giving a smaller positive or negative sum. When the two waves have equal values but opposite signs, they add to give zero.

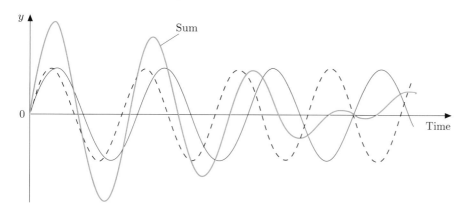

No vertical scale is shown in this figure: focus on the shape.

Figure 19   Adding two sine waves together. (Based on Figure 18.)

45

The sum of the two sine waves appears to be another periodic graph, though one whose amplitude also varies periodically. The amplitude reaches a peak when the 6 Hz and 7 Hz waves are in phase, and reduces to zero when the waves are out of phase. As the waves move in and out of phase every second, this pattern repeats itself over and over again.

▶   If this were an acoustic pattern, what would it sound like?

The amplitude of a sound wave determines the loudness of the sound. As you heard on the audio band, when two tones are very close together in frequency, you can hear a tone whose loudness regularly rises and falls, giving the distinctive rhythmic 'wah-wah-wah' sound called beats. In the case of the sum of the 6 Hz and 7 Hz waves in Figure 18(b), the beat pattern is repeated *once* every second. In general, the number of beats that occur each second is equal to the difference in frequency (in hertz) between the two tones.

You have now explored what happens in both *graphical* and *acoustic* terms when two sound waves are added together.

▶   But what is going on from an *algebraic* viewpoint? In other words, how can the sum of the sound waves be represented algebraically?

In Section 1.2, you saw that the formula for a sine wave can be expressed as $y = A \sin(2\pi f t)$, where A is the amplitude of the wave. If it is assumed in the present example that the amplitude of both the waves is 1, then the 6 Hz tone can be represented by $y = \sin 12\pi t$, and the 7 Hz tone by $y = \sin 14\pi t$. Hence, the effect of sounding these tones together is modelled by the sum

$$y = \sin 12\pi t + \sin 14\pi t. \tag{3}$$

This equation represents what is sometimes called a *beat waveform*.

You may like to try plotting on your calculator the three functions: $y = \sin 12\pi t$, $y = \sin 14\pi t$ and $y = \sin 12\pi t + \sin 14\pi t$. Use a time interval of 1 second.

3.3 Fourier's idea

The sounds made by musical instruments are much more complex than those in the case just considered but, for a sustained note, they are still periodic. Figure 20 presents two examples. In (a), you can see the waveform produced by a trumpet sounding the note A above middle C, and in (b), the waveform produced by a flute sounding the same note. Although the detailed shapes of the two waveforms are different, both are periodic and both have the same period. Because the frequency of A above middle C is 440 Hz for an instrument tuned to concert pitch (the conventional tuning of orchestral instruments), the period of both waveforms is $1/440 = 0.002273$ second, or about 2.3 milliseconds.

Recall that period $T = 1/f$.

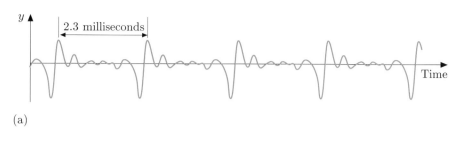

(a)

(b)

Figure 20   Waveforms produced by (a) a trumpet and (b) a flute.

▶   How are complex periodic waveforms like those in Figure 20 to be described?

In the early part of the nineteenth century, Jean-Baptiste Joseph Fourier developed a method, now known as Fourier analysis, that has enormous significance in many areas of mathematics, science and technology.

Historical note

Jean-Baptiste Joseph, Baron de Fourier (1768–1830) was involved in French revolutionary politics. Later, in 1798, he accompanied Napoleon on the Egyptian campaign, and subsequently became a leading civil administrator and secretary to the Egyptian Institute, which Napoleon had founded. In 1801, he returned to France with the responsibility of publishing the enormous quantity of research on Egyptian antiquities carried out by the Institute. He became Préfet (senior administrator) of the Department of Isère at Grenoble, and during this time he also worked on problems in mathematical physics. In 1822, he published his work on the theory of heat, where he used combinations of sine and cosine functions to model heat flow.

The basis of Fourier analysis is that *any* periodic curve can be treated as if it were made up of sine curves with frequencies that are related in a particular way. As a starting point, consider a periodic—but somewhat unmusical—waveform: the *square wave*. This general shape is shown in Figure 21 overleaf. Waveforms like this turn up in communications, digital systems, and the electronic production of sounds. At first sight, it seems very unlikely that such a shape could be made up of sine curves.

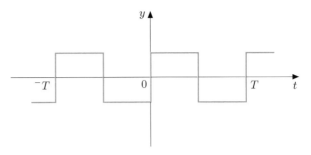

Figure 21   A square wave of period T.

▶   How would you describe a square wave mathematically?

A square wave can be approximated by adding together a specific choice of sine curves. Look at Figure 22, which shows how such a wave is built up.

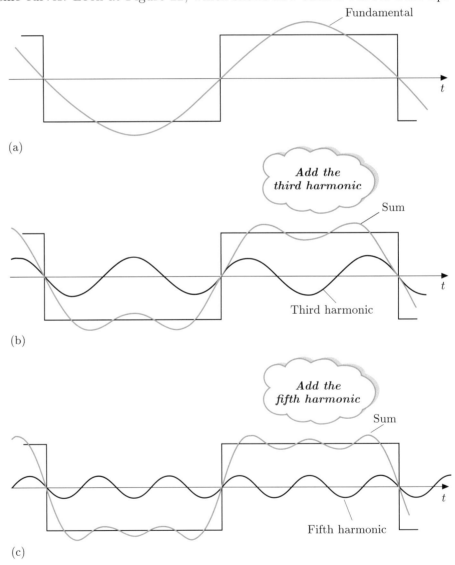

Figure 22   Approximating a square wave by (a) the fundamental, (b) adding the third harmonic, (c) adding the fifth harmonic.

In Figure 22(a), a single sine wave with the same period T as the square wave offers the first approximation. The frequency, $f = 1/T$, of this sine wave is called the *fundamental* frequency. The sine wave itself is called the *fundamental component*; it is also known as the *first harmonic*. In general, the term 'harmonic' is used to refer to a sound component with a frequency that is an integer multiple of the fundamental. (You may recall that in the video you heard the students in the music laboratory talk about the harmonics associated with the fundamental frequency of a musical note.)

Figure 22(b) shows the sum of the fundamental component and a sine wave with a frequency that has been chosen to be three times the fundamental frequency. (You will see the effect of this particular choice shortly.) The added component is called the *third harmonic*. Notice that the amplitude of this new component is less than that of the fundamental. The sum of the fundamental and the third harmonic produces a waveform that has the same period as the square wave, and is a better approximation to it than the fundamental alone.

In Figure 22(c), the sum obtained by adding a further sine wave is shown. This time the frequency of the new component has been chosen to be five times that of the fundamental. Its amplitude is smaller than that of either of the other two components. Adding this *fifth harmonic* to the sum of the fundamental and the third harmonic improves the approximation to the square wave even more.

The process does not have to stop with the fifth harmonic. If you keep on adding harmonic sine wave components of increasing frequency and diminishing amplitude, you will continue to improve the approximation to the square wave. The mathematical details are beyond the scope of this course, but the general interpretation of the results is relatively straightforward. The square wave can be thought of as being made up of the sum of a fundamental sine wave with a frequency f (where $f = 1/T$, and T is the period of the square wave), and harmonic sine waves with frequencies $3f$, $5f$, $7f$, $9f$, and so on. The amplitude of each harmonic component of the square wave turns out to be inversely proportional to the component's frequency: thus the amplitude of the component with frequency $3f$ is $1/3$, with $5f$ it is $1/5$, and so on.

The end result is that the square wave can be represented by the following sum of sine waves:

$$y = \sin(2\pi f t) + \frac{1}{3}\sin[2\pi(3f)t] + \frac{1}{5}\sin[2\pi(5f)t] + \frac{1}{7}\sin[2\pi(7f)t] + \cdots,$$

with each additional term improving the approximation.

If angular frequency is used rather than frequency, you can replace $2\pi f$ by ω, $2\pi(3f)$ by 3ω, and so on, and write the sum of the sine waves as

$$y = \sin \omega t + \frac{1}{3}\sin 3\omega t + \frac{1}{5}\sin 5\omega t + \frac{1}{7}\sin 7\omega t + \cdots.$$

This sum is called a *Fourier series*.

If you go on to higher-level courses in mathematics, technology or science, you may meet Fourier series again.

Example 4   Building up a square wave

A given square wave has a frequency of 1 Hz.

(a) What are the frequencies, in hertz, of the fundamental and the first few harmonic components (third, fifth, seventh, and so on) of the wave?

(b) Write down the Fourier series corresponding to this square wave.

Solution

(a) The frequency of the fundamental has to be the same as the frequency of the square wave; that is, 1 Hz. The frequencies of the harmonics are odd multiples of the fundamental: $3 \times 1 = 3\,\text{Hz}$, $5 \times 1 = 5\,\text{Hz}$, $7 \times 1 = 7\,\text{Hz}$, and so on.

(b) The Fourier series for the square wave is

$$y = \sin 2\pi t + \frac{1}{3}\sin 6\pi t + \frac{1}{5}\sin 10\pi t + \frac{1}{7}\sin 14\pi t + \ldots .$$

The components 2π, 6π, 10π, 14π and so on are the angular frequencies of the fundamental, the third harmonic, the fifth harmonic, the seventh harmonic and so on, expressed in radians per second.

Activity 21   Components of a square wave

Another square wave has a period of 0.5 second.

(a) What are the frequencies, in hertz, of the fundamental and the first few harmonic components (third, fifth, seventh, and so on) of the wave?

(b) Write down the Fourier series corresponding to this square wave.

(c) Use your calculator to display the fundamental component, then add in the next component and display the result, and so on for the first four components. Plot values from 0 to $\pi/3$ on the X-axis. What shape do you see building up?

A different mix of harmonic components can change the wave shape significantly. A waveform known as a sawtooth wave, for example, uses the fundamental and *all* the harmonics (not just the odd-numbered ones) and has the Fourier series

$$y = \sin \omega t + \frac{1}{2}\sin 2\omega t + \frac{1}{3}\sin 3\omega t + \frac{1}{4}\sin 4\omega t + \frac{1}{5}\sin 5\omega t + \cdots .$$

As with the square wave, each new term improves the approximation to the original wave.

Activity 22 *Components of a sawtooth wave*

What is the period of a sawtooth wave if the value of the fundamental angular frequency ω is 20 radians per second? Write down the Fourier series corresponding to this waveform.

Use your calculator to display the waveform you obtain when using the first four terms of the Fourier series.

In Section 2.2, you saw that shifting the phase of a sine wave by π radians (180°) effectively turned the curve upside down. This result can be extended to the Fourier series of a periodic waveform. If you shift the phase of each sine wave component in a Fourier series by π radians, then the result is the same as turning the entire waveform upside down. For example, a sawtooth waveform can be inverted by shifting each component in its Fourier series by π. This gives the Fourier series for the inverted wave:

$$y = \sin(\omega t + \pi) + \frac{1}{2}\sin(2\omega t + \pi) + \frac{1}{3}\sin(3\omega t + \pi) + \frac{1}{4}\sin(4\omega t + \pi) + \cdots.$$

Multiplying any sine function by $^-1$, in fact, gives the same result as a phase shift of π. This leads to the identity:

$$\sin(x + \pi) = {}^-\sin x.$$

If you apply this result to the equation for the inverted sawtooth wave, you get an alternative equation for that wave:

$$y = {}^-\sin\omega t - \frac{1}{2}\sin 2\omega t - \frac{1}{3}\sin 3\omega t - \frac{1}{4}\sin 4\omega t - \cdots$$

or

$$y = {}^-\left(\sin\omega t + \frac{1}{2}\sin 2\omega t + \frac{1}{3}\sin 3\omega t + \frac{1}{4}\sin 4\omega t + \cdots\right).$$

This last equation is actually the original series multiplied by $^-1$. If you add this series to the original series, the sine terms will cancel each other out and you will be left with nothing. When the Fourier series represents a sound wave, the result of adding two sounds whose frequency components are phase-shifted by π radians relative to each other will be silence.

Activity 23 *Turn off the sound*

On your calculator, display the sawtooth wave from Activity 22. Then enter the Fourier series for the sawtooth wave, this time adding a phase shift of π radians to each sine wave component. Display the original and the new waveforms together. What do you see?

Finally, display the sum of the two waveforms. What do you now see?

Now work through Section 15.3 of the Calculator Book.

This section of the *Calculator Book* will help you to create a program that will produce various waveforms by using Fourier series. If you have time, also work through the optional Section 15.4, which explains some curious features of your calculator's display. These have important implications for sampling and aliasing—two key ideas in digital processing.

3.4 The frequency spectrum

The Fourier series offers a new way of thinking about periodic waveforms because any such waveform can be expressed as a sum of harmonic frequency components. Furthermore, this sum is unique: any given waveform is associated with one and only one sum of harmonic components. The range of frequency components that characterizes a specific waveform is called its *frequency spectrum*.

Light is an electromagnetic wave with some properties very much like those of the waves discussed in Section 3.3.

You will learn more about the mathematics of the rainbow in *Unit 16*.

You may well have come across the idea of a spectrum in other contexts. For example, the visible spectrum is the range of light frequencies, corresponding to colours from red to violet, to which our eyes respond. Sunlight can be thought of as being largely made up of the visible spectrum; thus, in a rainbow you see the effect of droplets of water splitting sunlight into the colours of the visible spectrum—from the lower-frequency red light at one end, to the higher-frequency violet light at the other.

At its most basic, a frequency spectrum is a range of frequencies. A graphical way of representing such a spectrum is shown in Figure 23. The horizontal axis gives the frequency, and the vertical axis gives the amplitude. On this plot, a single sine wave with an amplitude A and a frequency f is represented by a vertical line of height A at a position f on the frequency axis.

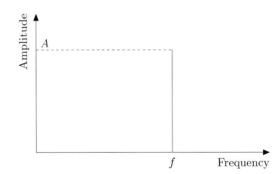

Figure 23   Frequency spectrum of a single sine wave.

A more complex wave will have a more complex spectrum. In Section 3.2, you saw how beats are formed when two tones whose frequencies differ by just a few hertz are added together. Each tone was modelled as a sine wave, and each sine wave was a frequency component of the beat waveform. This was illustrated in Figure 18 for tones with frequencies of 6 Hz and 7 Hz, and the resulting beat waveform was modelled by equation (3). Figure 24 shows the corresponding frequency spectrum. In

SECTION 3   PATTERNS OF SOUND

it, there are two lines, one associated with the 6 Hz sine wave and the other with the 7 Hz sine wave. The heights of the lines are equal, showing that the amplitudes of the sine wave components are equal.

Note that on a plot of a frequency spectrum, the lower frequencies lie to the left and the higher frequencies lie to the right.

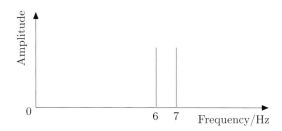

Figure 24   Frequency spectrum of a beat waveform produced by tones of frequencies 6 Hz and 7 Hz.

In Figure 25, part of the frequency spectrum of a square wave is shown. It emphasizes the fact that such a wave is made up of several components. The line corresponding to the fundamental frequency has a height of 1 at a frequency f. The harmonic components at frequencies $3f$, $5f$, $7f$, $9f$, and so on, are represented by lines with heights 1/3, 1/5, 1/7, 1/9, respectively. The amplitudes of the high-frequency components of the square wave get smaller as the frequency increases, but they never vanish completely.

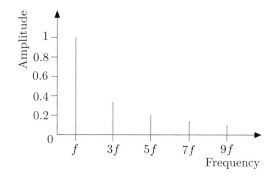

Figure 25   Part of the frequency spectrum of a square wave.

As you saw in the video, the important point about the frequency spectrum of a periodic waveform is that each vertical line represents a separate sine wave. The position of the line along the horizontal axis indicates the frequency, and its height shows the amplitude of that sine wave.

To recap, a periodic waveform can be represented either by a graph of its amplitude plotted against time, or by a frequency spectrum. They are complementary representations; each stresses some features of the waveform but ignores others. The graph shows the value or size of the wave at each instant in time, but gives no information about how the frequency components of the waveform are distributed. On the other hand, the spectrum gives explicit information about the frequency components but little hint about the shape of the waveform. The two representations are like the two sides of a coin; neither tells the whole story, but together they give two useful views of what is going on.

Activity 24 *Sketching the spectrum*

Look back to Section 3.3 to find the Fourier series for the square wave and for the sawtooth wave. Using this information, sketch the frequency spectrum of each of the following periodic waves for frequencies up to 10 000 Hz (10 kHz):

(a) a square wave with a frequency of 1 kHz;

(b) a sawtooth wave with a period of 1 millisecond (10^{-3} seconds).

In each case, assume that the amplitude of the fundamental is 1.

A frequency spectrum for a musical note represents one moment in time. Clearly, if the note swells or dies away, the picture changes over time. But how can such changes be represented graphically? Since a spectrum is a two-dimensional representation, it can show only two variables at once, but there are three variables here—frequency, amplitude and time. Changes over time can be depicted by ignoring the frequency components and plotting amplitude against time.

You have now seen three ways of representing complex periodic variations:

- a graph showing the waveform, with amplitude plotted against time;
- an algebraic representation in the form of a Fourier series;
- a frequency spectrum, with amplitude plotted against frequency.

It is useful to consider how the two diagrammatic representations—the graph of the waveform itself, and the frequency spectrum—give different insights into the sounds produced by musical instruments. This will be explored in the case of the trumpet and the flute. Figure 26 shows the waveforms produced by these instruments sounding the note A at 440 Hz.

You have already seen these waveforms in Figure 20.

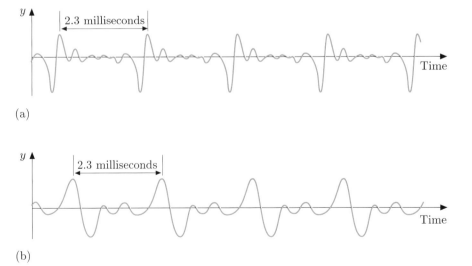

(a)

(b)

Figure 26 Waveforms produced by (a) a trumpet and (b) a flute.

▶   What are the similarities and the differences between the two
     waveforms?

Both waveforms have the same period of 1/440 second, or about
2.3 milliseconds. The flute has a characteristically soft and pure tone, and
this corresponds to a relatively smoothly-changing waveform, (b). During
one period the waveform goes through one large oscillation followed by two
smaller ones. There are no sudden changes, and the turning points in the
waveform are smooth and rounded. In contrast, the trumpet produces a
sound sometimes described as 'bright' or 'brassy'. Its waveform, (a), also
has one large oscillation during each period, but this is followed by five
smaller peaks. The trumpet's waveform changes faster (the slopes are
steeper) than the waveform of the flute, and the turning points are
sharper, especially during the largest oscillation.

▶   How are these qualitative differences related to the quality, or timbre,
     of the sounds of the flute and the trumpet?

The frequency spectrum, rather than the waveform, provides the more
helpful picture in understanding the perceived quality of a sound. In
Figure 27, you can see the frequency spectrum diagrams corresponding to
the two waveforms in Figure 26.

These diagrams are similar to
those you saw being produced
in the video.

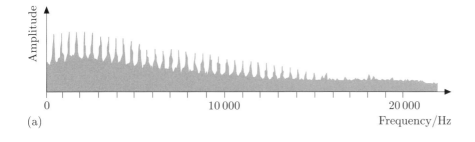

(a)

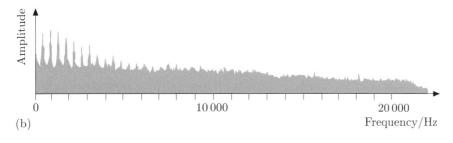

(b)

Figure 27   The frequency spectrum of (a) a trumpet and (b) a flute.

The frequency scale on the horizontal axis in Figure 27 is from 0 Hz to over
20 000 Hz (20 kHz). The two frequency spectrum diagrams appear as a
series of peaks or spikes over this range, but they are actually made up of a
very large number of vertical lines. Each line represents a sine wave
component of a particular frequency, and the height of the line represents
the amplitude of that component.

The human ear is sensitive to
frequencies from about 20 Hz
to about 20 kHz. This range
becomes narrower as people
get older.

In these spectra, the fundamental is not the component of maximum amplitude.

Since the trumpet and the flute are sounding the same note, each spectrum shows the same fundamental frequency component at 440 Hz. This is represented by the first spike in each spectrum. The spikes following the fundamental represent the higher-frequency harmonics of the sound.

Notice the difference between the range of frequencies produced by the two instruments. The harmonics of the flute extend up to about 5000 Hz, after which it is difficult to distinguish them clearly. The harmonics of the trumpet, however, extend up to at least 15 000 Hz, three times further than in the case of the flute. Notice also that the height of the spikes in the trumpet's spectrum decreases at a much slower rate than for the flute. Unlike the flute, the trumpet is a rich source of strong harmonics over a wide range of frequencies.

The differences between the frequency spectrum of the trumpet and that of the flute give direct insight into the quality of the sound produced by each instrument. The trumpet, with its complex 'brassy' sound, has a spectrum with numerous harmonics, whereas the relatively pure note of the flute is characterized by a spectrum with comparatively few harmonic components.

What you should take away from this discussion is a broader view of periodic behaviour and an appreciation that not all the information about a given situation will be contained in a useful way in just one type of graphical representation or just one mathematical formula. To describe some aspect of reality, you may need several different models—each one giving you a different frame within which to picture the world.

If you have time, you should now watch the video 'The sound of silence' again, and complete the following activity.

Activity 25   *Explaining about sounds*

While watching the video, you should pay special attention to the parts concerning the representation of more-complex sounds. As you watch, jot down points to help you with the rest of the activity, starting and stopping the DVD as necessary.

Using ideas from the video and the text, make some notes in order to explain the following to someone who is not doing this course.

(a) What is meant by the term 'frequency spectrum'?

(b) Why do the sounds made by a trumpet and a flute both playing the same note have different qualities?

Section 3 has looked at more complex periodic waveforms. As you have seen, these can be thought of as being made up of sine wave components. A plot of the amplitudes of these components against their frequencies gives a frequency spectrum representation of the waveform. Periodic waveforms can also be represented by plotting a graph of their amplitude against time. The two representations are complementary. Each stresses aspects of the behaviour that the other ignores.

Outcomes

After studying this section, you should be able to:

◊ use the following terms accurately and explain them: 'sine wave', 'sinusoid', 'waveform', 'beat', 'fundamental', 'harmonic', 'Fourier series', 'frequency spectrum' (Activities 19, 20, 21, 22, 24 and 25);

◊ appreciate that any periodic waveform can be represented by a sum of sine waves or a frequency spectrum, and use and interpret such representations (Activities 21, 22, 24 and 25);

◊ use your calculator to explore how periodic waveforms can be built up from sine functions using Fourier series (Activities 21, 22 and 23).

Unit summary and outcomes

This unit has been about using sine waves to model periodic behaviour.

Section 1 reviewed some of the mathematical language of sine curves. The amplitude gives the maximum height of a sine curve, above or below the horizontal axis, and the period gives the width of the curve—the time for one complete cycle. You saw how a sine curve can be generated by circular motion, as exemplified by an exercise bike. Plotting the height of one of the bike's pedals, relative to the pivot, against time gives a sine curve whose amplitude is equal to the radius of the circle described by the pedal about the pivot; the period is equal to the time taken for one complete revolution. The section ended with a look at some simple transformations based on the graph $y = \sin x$.

Section 2 developed a sine wave model of the change in the sunset times over the course of a year. You saw how the mathematical curve was matched to the sunset data by choosing appropriate values of amplitude, frequency (or period), phase shift and an additive constant. The model was tested by comparing its predictions with the actual data. Also in this section you met some trigonometric identities that were derived by considering particular phase shifts of sine and cosine curves.

Section 3 looked at some periodic sound patterns. Pure tones sounded together can produce beats, or rhythmic variations in amplitude. This section showed how beats are produced and how they can be modelled mathematically. The video explored the question of why different musical instruments playing the same note sound different. To answer this question, two graphical representations of a musical note were used—the waveform and the frequency spectrum. In the audio section you heard how piano tuner Henry Tracy used the phenomenon of beats to tune a piano according to the principle of equal temperament. For more complex periodic patterns, Fourier analysis shows how sine waves can be used as building blocks. Each sine wave can be thought of as a frequency component of the waveform. A periodic waveform can then be represented by a frequency spectrum showing these components. A waveform and its associated spectrum are complementary graphical representations; each stresses aspects that the other ignores.

Outcomes

Now that you have finished this unit you should be able to:

◇ use the following terms accurately and explain them: 'periodic behaviour', 'sine wave', 'sinusoid', 'amplitude', 'period', 'frequency', 'angular frequency', 'cycles per second', 'hertz', 'radian', 'phase shift', 'identity', 'beat', 'waveform', 'fundamental', 'harmonic', 'Fourier series', 'frequency spectrum';

◇ describe in your own words the relationship between circular motion and a sine curve;

◇ explain and use the mathematical relationships between frequency, angular frequency and period;

◇ write down the formula for a sine curve, given the amplitude and either the period, frequency or angular frequency;

◇ outline the modelling process involved in fitting a sine curve to a set of periodic data;

◇ explain how to translate a sine curve or cosine curve vertically or horizontally, by adding an appropriate constant or a phase shift;

◇ relate the sine and cosine functions to each other, using standard trigonometric identities;

◇ interpret and evaluate an expression of the form $M + A\sin(\omega t + \phi)$, and use it in modelling;

◇ appreciate that any periodic waveform can be represented by a sum of sine waves or a frequency spectrum, and use and interpret such representations.

With the aid of your calculator, you should also be able to:

◇ model data using sine regression;

◇ choose appropriate values for the amplitude, period (or frequency) and phase shift of a sine curve model, given a set of periodic data;

◇ explore how periodic waveforms can be built up from sine functions using Fourier series.

Comments on Activities

Activity 1

There are no comments for this activity.

Activity 2

The parameters of the sine waves are:

(a) Amplitude: 1.
Period: 1 s.
Frequency: 1 Hz.

(b) Amplitude: 0.5.
Period: 4 s.
Frequency: 0.25 Hz.

(c) Amplitude: 50.
Period: 0.02 s.
Frequency: 50 Hz.

Activity 3

A quarter of a turn (90°) corresponds to $\frac{1}{4} \times 2\pi = \pi/2$ radians. Half a turn (180°) is $1/2 \times 2\pi = \pi$ radians, and three-quarters of a turn (270°) is $3/4 \times 2\pi = 3\pi/2$ radians.

Activity 4

(a) The formula describing the sine curve is $y = 0.1 \sin(2\pi \times 3 \times t)$, or $y = 0.1 \sin(6\pi t)$.

(b) The period T of the curve is the reciprocal of the frequency in hertz, so $T = 1/3$ s.

(c) The first peak of the sine curve occurs a quarter of a period after $t = 0$. Since the period is $1/3$ s, the peak occurs at $t = (1/3)/4 = 1/12$ s, or 0.0833 s.

Your sketch should look something like Figure 28.

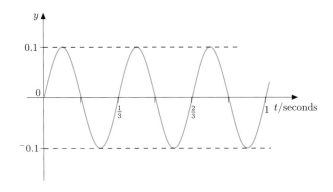

Figure 28

Activity 5

(a) Since the period T is required, the equation $y = 0.001 \sin(30t)$ can be compared with the general formula which includes T:

$$y = A \sin\left(\frac{2\pi}{T} t\right),$$

where A is the amplitude.

It follows that

$$A = 0.001.$$

Also

$$\frac{2\pi}{T} = 30,$$

so

$$T = \frac{2\pi}{30} = 0.21 \text{ s (to 2 d.p.)}.$$

(b) Enter the equation as $Y1 = 0.001 \sin(30X)$ on your calculator. Choose appropriate window settings that will accommodate the values of the amplitude and the period; for example, X from 0 to 0.5 and Y from $^-0.001$ to $+0.001$.

Two complete cycles will take 0.42 s.

(c) Since the frequency f is given, it is easiest to use $y = A \sin(2\pi f T)$. Here $f = 8$, so the formula becomes

$$y = 0.001 \sin(16\pi t).$$

Enter $Y2 = 0.001 \sin(16\pi X)$ on your calculator. The new curve has a shorter period (0.125 s), with over three cycles displayed, corresponding to the first curve's two cycles.

Activity 6

Read the summary that appears before the activity and check that you have included all the points.

Activity 7

In general, the function $y = \sin(x)$ is transformed to $y = a + b \sin(cx + d)$ as follows:

a shift upwards of amount a,
a vertical scaling with a factor b,
a horizontal 'squashing up' with a factor c,
a shift to the left of amount d.

Activity 8

The period T is one year, or 52 weeks.

From midsummer to midwinter, the time of sunset varies over a total range of about 4.5 hours. A suitable value for the amplitude is, therefore, half this value, 2.25 hours.

Activity 9

Ensure that your calculator is set to radian mode. Check your graph against Figure 10, using the following window settings: $X\text{min} = {}^-8$, $X\text{max} = 56$, $Y\text{min} = 15$, $Y\text{max} = 21$.

Activity 10

(a) Over the interval $0 \leq x \leq 2\pi$, the value of $\sin(x + \pi/2)$ is equal to 0 for $x = \pi/2$ and $x = 3\pi/2$. The function is equal to $^-1$ for $x = \pi$, and equal to $+1$ for $x = 0$ and $x = 2\pi$.

(b) The curve $y = \sin x$ must be shifted to the left by a phase shift of $\pi/2$ radians in order to coincide with $y = \sin(x + \pi/2)$.

Activity 11

The improved model is represented by

$$y = 18.12 + 2.25 \sin\left(\frac{2\pi}{52}t - 1.21\right).$$

Substituting $t = 43$ gives

$$\begin{aligned}
\text{sunset time} &= 18.12 \\
&\quad + 2.25 \sin\left(\frac{2\pi}{52} \times 43 - 1.21\right) \\
&= 18.12 + 2.25 \sin(5.20 - 1.21) \\
&= 18.12 + 2.25 \sin(3.99) \\
&= 18.12 - 1.68 \\
&= 16.44, \text{ or } 16.26 \,\text{GMT}.
\end{aligned}$$

The actual sunset time for the start of the week in which Guy Fawkes' Night falls is 16.31 GMT. In this case the difference between the actual value and the time predicted by the improved model is only 5 minutes.

Activity 12

A key feature of the modelling process is that you start with a simple model and gradually refine it.

Here are some of the assumptions that have been made. One is that the sunset data are essentially discrete—the Sun sets only once each day. There is no 'sunset curve' (even if based on daily, rather than weekly data), only a set of numbers. Plotting these data and then 'seeing' a curve requires making a modelling assumption. The shape of the curve generated by the discrete data is similar to that of a continuous sine curve, and so a sine function is used in the model. However, the model should not be used to predict the time of sunset for *any* value of t, but just for the integer values $0, 1, 2, \ldots, 51$ and those such as $\frac{1}{7}, \frac{2}{7}, \frac{3}{7}, \ldots$ representing intermediate days.

The model also assumes that the shape of the sunset curve will always be the same. This means that the model might not be useful for making predictions over thousands of years, because it ignores gradual changes in the Earth's orbit.

What constitutes a good fit in a model depends on what you want to use the model for, and how numerically accurate you want it to be. As you have seen in previous units, there are various mathematical techniques that define goodness of fit. These work to adjust the parameters of the proposed model so that the 'best' fit using a particular criterion is found. Depending on your requirements, you may want a model whose predictions do not fall outside a certain range, or, alternatively, where the average error is least.

Activity 13

The cosine function produces a curve that is identical to the sine curve—*except* that it is shifted in phase (relative to the sine curve) by a quarter of a period, or $\pi/2$ radians, to the left.

The effect of this phase shift is that the cosine curve starts at $y = 1$ at $x = 0$, instead of at $y = 0$. Otherwise the sine and cosine curves share the same periodic shape, the same amplitude (equal to 1), and the same period (2π radians).

Activity 14

The $\pi/2$ phase shift effectively shifts the $\cos x$ curve to the left along the x-axis. As a result, the $\cos(x + \pi/2)$ curve is identical to the curve of $^-\sin x$, and you can write

$$\cos(x + \pi/2) = ^-\sin x.$$

You can check this identity for particular values of x. For example, if $x = 0$, then $\cos(\pi/2) = 0$, which is the same as $^-\sin(0) = 0$. If $x = 1$, then $\cos(1 + \pi/2) = \cos(2.5708) = ^-0.8415$, which is the same as $^-\sin(1) = ^-0.8415$. Try some other values of x for yourself.

Note that, because of the cyclical nature of the sine and cosine curves, there are infinitely many possible phase shifts that will produce this result, so you might have a different identity; for example, $\cos(x + \pi/2) = \sin(x + \pi)$ or $\sin(x - \pi)$.

Activity 15

Expression	Simpler form	Comment
$\sin(x - \pi/2)$	$^-\cos x$	$\sin x$ has been shifted to the right by $\pi/2$ radians
$\cos(x + 2\pi)$	$\cos x$	$\cos x$ has been shifted to the left by 2π radians
$\sin(x - \pi)$	$^-\sin x$	$\sin x$ has been shifted to the right by π radians
$\cos(x + 3\pi/2)$	$\sin x$	$\cos x$ has been shifted to the left by $3\pi/2$ radians

Activity 16

Since $\sin x = \cos(x + 3\pi/2)$, the model can be rewritten as

$$y = 18.12 + 2.25 \cos\left(\frac{2\pi}{52}t - 1.21 + \frac{3\pi}{2}\right)$$

$$= 18.12 + 2.25 \cos\left(\frac{2\pi}{52}t + 3.50\right).$$

Check for yourself that the two versions of the model give the same results (to 2 d.p.) for different values of t.

There are many alternatives to the formula above. For example, using the identity $\sin(x) = \cos(x - \pi/2)$ gives

$$y = 18.12 + 2.25 \cos(\frac{2\pi}{52}t - 2.78).$$

Activity 17

Over any range of values of θ, the sum of $\sin^2 \theta$ and $\cos^2 \theta$ is a horizontal straight line of constant height 1. In other words, $\sin^2 \theta + \cos^2 \theta = 1$.

Activity 18

There are no comments for this activity.

Activity 19

(a) As a point moves up and down, its vertical displacement (that is, its vertical height above the horizontal), when plotted against time, traces out a sine curve. The displacement is the *amplitude* of the sine curve. The time to go from the horizontal, through a maximum displacement, then to a minimum displacement and return to the horizontal corresponds to one cycle of the sine function: this is the *period*. The *frequency* is the number of cycles completed in unit time; this corresponds to the frequency of oscillation of the sine curve. The *phase shift* corresponds to where in the cycle the oscillation starts; it is a measure of the horizontal translation of the sine curve from the standard sine curve.

Complex patterns formed by adding different sine curves together can be represented graphically either as a *waveform*, where amplitude is plotted against time, or as a *frequency spectrum*, where the amplitude of each component sine wave is plotted against its frequency.

(b) The 'trick' is to invert the waveform of one note and then add it to the waveform of the original note. The two waveforms then cancel one another out, producing silence.

Activity 20

Henry starts by blocking off two of the strings that make up the note C above middle C. He then tunes the third central string to exactly the pitch of the tuning fork. Next he tunes middle C in the same way.

He then blocks off the two outside strings of every other note in the bearing scale, and tunes the centre string of each.

Next Henry goes up a fourth from middle C to F above middle C, and then from F down a fifth to B♭, then up a fourth to E♭, and so on. The only variation is where he goes up two fourths in succession in order to stay within the octave of the bearing scale.

Henry next tunes the unisons, which are the other two strings that make up any note in the bearing scale. He then transfers this bearing scale up and down the piano, copying a full octave note by note, before moving on to the next octave.

Activity 21

(a) The frequency of the fundamental is the same as the frequency of the square wave, so $f = 1/0.5 = 2$ Hz. The frequencies of the harmonics are odd multiples of the fundamental: $3 \times 2 = 6$ Hz, $5 \times 2 = 10$ Hz, $7 \times 2 = 14$Hz, and so on.

(b) The Fourier series for the square wave is

$$y = \sin 4\pi t + \frac{1}{3}\sin 12\pi t + \frac{1}{5}\sin 20\pi t$$
$$+ \frac{1}{7}\sin 28\pi t + \cdots .$$

(c) You should see the waveform becoming less like a sine wave and more like a square wave.

Activity 22

The period T of the sawtooth wave is the same as that of the fundamental, so

$$T = 2\pi/\omega = 2\pi/20 = 0.3142\,\text{s}.$$

The Fourier series is

$$y = \sin 20t + \frac{1}{2}\sin 40t + \frac{1}{3}\sin 60t$$
$$+ \frac{1}{4}\sin 80t + \frac{1}{5}\sin 100t + \cdots .$$

Activity 23

The Fourier series for the phase-shifted sawtooth wave is

$$y = \sin(20t + \pi) + \frac{1}{2}\sin(40t + \pi) + \frac{1}{3}\sin(60t + \pi)$$
$$+ \frac{1}{4}\sin(80t + \pi) + \cdots.$$

When you display this series, you should see the shape of the original sawtooth wave, but upside down.

The result of adding the original and the phase-shifted waveforms is $y = 0$, a straight line along the x-axis.

Activity 24

(a) A square wave with a frequency of 1000 Hz (or 1 kHz) will have a fundamental component at 1 kHz and harmonics at odd multiples of the fundamental: at 3 kHz, 5 kHz, 7 kHz, 9 kHz, and so on.

Figure 29(a) shows the frequency spectrum of the square wave up to 10 kHz (10 000 Hz). Notice how the amplitude of the harmonic frequency components gets smaller as the frequency increases.

(b) The fundamental frequency of the sawtooth wave will be $1/0.001 = 1000\,\text{Hz} = 1\,\text{kHz}$. The sawtooth wave has frequency components at the fundamental and its integer multiples.

Figure 29(b) shows the frequency spectrum. As with the square wave, the amplitudes of the frequency components decrease with increasing frequency, but they never completely disappear.

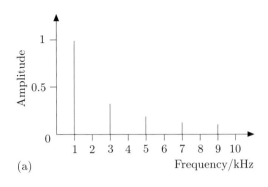

(a)

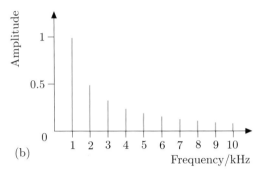

(b)

Figure 29   Frequency spectra for (a) a square wave and (b) a sawtooth wave.

Activity 25

Everybody's notes will be different, but you should mention that in a frequency spectrum, the amplitudes of the different frequency components of the sound are plotted, and that the frequency spectrum of the trumpet contains more high amplitude harmonics than that of the flute, hence they sound different.

Acknowledgements

Cover

Indian rock art rainbows: Kenneth Sassen and Optical Society of America; other photographs: Mike Levers, Photographic Department, The Open University.

Index